WHAT EVERY ENGINEER SHOULD KNOW ABOUT

ACCOUNTING AND FINANCE

WHAT EVERY ENGINEER SHOULD KNOW

A Series

Editor

William H. Middendorf

Department of Electrical and Computer Engineering
University of Cincinnati
Cincinnati, Ohio

WHAT EVERY ENGINEER SHOULD KNOW ABOUT

ACCOUNTING AND FINANCE

Jae K. Shim
California State University, Long Beach
Long Beach, California

Norman Henteleff
Consultant in Engineering Management
and
National Business Review Foundation
Long Beach, California

Marcel Dekker, Inc.   New York • Basel • Hong Kong

Library of Congress Cataloging-in-Publication Data

Shim, Jae K.
    What every engineer should know about accounting and finance / Jae
K. Shim, Norman Henteleff.
        p.   cm. — (What every engineer should know ; v. 32)
    Includes index.
    ISBN 0-8247-9271-8
    1. Engineering—Accounting. 2. Engineering firms—Finance.
I. Henteleff, Norman. II. Title. III. Series.
TA185.K56   1994
620.00681—dc20   94-29763
                                                                CIP

The publisher offers discounts on this book when ordered in bulk quantities. For more information, write to Special Sales/Professional Marketing at the address below.

This book is printed on acid-free paper.

MARCEL DEKKER, INC.
270 Madison Avenue, New York, New York 10016

Current printing (last digit):
10 9 8 7 6 5 4 3 2 1

PRINTED IN THE UNITED STATES OF AMERICA

PREFACE

This book is directed toward the engineer who must have accounting and financial knowledge but has not had formal training in accounting or finance—perhaps a newly promoted middle engineering manager or a project manager who must have this knowledge to perform engineering and project analyses and to communicate with the financial officers in various areas of his or her function. This book is also targeted to engineers who aspire to be entrepreneurs or sole proprietors; they may have brilliant product ideas, but not the slightest idea about financial management, including financing. This book will ease the transition from effective engineer to successful engineering manager. The book's goals are threefold:

1. To provide an understanding and working knowledge of the fundamentals of finance and accounting that can be put to practical application in day-to-day jobs of engineers and engineering managers.
2. To provide a working vocabulary for communication, so that the engineer can develop an ability to ask the right questions and interpret jargon-based financial answers.
3. To cover a variety of methods, processes, and tools of accounting and finance with many examples and illustrations drawn from the fields of engineering and project management.

Project managers and engineers cannot avoid financial information. Profitability statements, rates of return, budgets, variances, and project analyses are included in the engineer's job. The financial aspects of operations are a part of the project manager's day-to-day considerations.

Most engineers who are promoted to management positions have years of training and experience in their technical areas but very limited education or experience in administrative or project management.The book will enable engineers and project managers to prepare, appraise, evaluate, and approve plans to accomplish departmental and company objectives. It will also enable them to prepare financial support documents that must accompany a given program or project.

The transition from civil engineer, chemical engineer, electrical engineer, mechanical engineer, industrial engineer, systems engineer, radio/television engineer, or scientist to engineering or project manager can be, at the least, unnerving or unsettling. For new engineering managers, the first item of business is survival. They must be knowledgeable and able to communicate with lawyers, accountants, financial officers, quality specialists, other engineers, and a host of upper and middle management people.

Engineering managers must be able to express budgetary needs in order to obtain proper funding for their departments. They may have to forecast future sales, cash flows, and costs to determine whether their company will be operating effectively in the future. A knowledge of variance analysis will help the project manager spot areas of inefficiency or efficiency by comparing actual performances to standards. An understanding of certain financial strategies will help the engineering manager to improve return on investment and enhance profitability as well as use the company's assets more effectively. The concepts of break-even analysis and the time value of money are excellent financial tools enabling the project manager to choose whether to buy or lease, or purchase machine A or machine B, thus improving the investment opportunities of the company. Cost analysis can be applied to appraising and evaluating costs associated with a decision or project, including effects on profitability.

Understanding the concepts and conventions of accounting and finance will assist engineers in the proper interpretation of financial data. For example, knowing how facilities are depreciated is a necessity for the engineer when making investment decisions since depreciation expenses directly affect taxable income, and tax considerations often prove to be a deciding factor. Financial concepts such as debit and credit, accrual versus cash basis, and return-risk tradeoff need to be understood for proper interpretation and analysis of accounting data.

Understanding financial statement analysis gives the engineering manager a working knowledge of how to evaluate the company's health and operating performance, and how to investigate trends in profitability and return on investment.

For the engineer or engineering manager the budget offers a means to do a better job and to prevent failure. Reporting performance against budget goals and standards permits measurement and control of engineering activities. Cost reporting leads to better definitions of objectives, more effective evaluation of work in progress, and better measurement of performance at every level. Cost reporting is an important phase of measuring and evaluating engineering projects.

The knowledgeable project manager has a number of methods at hand to choose among new product developments or alternative facilities. These include discounted cash flow (DCF) analysis such as uniform annual cost, net present value, internal rate of return, payback period, and scoring and ranking methods.

Often the engineering manager has to evaluate time-cost tradeoffs, for example, whether to increase resource allocation to specific tasks so that a project can be finished at an earlier date.

The focus of this book is to create an awareness and understanding of specific finance and accounting processes, methods, strategies (such as tax advantages of leasing), and terminology (such as depreciation, leverage, financial ratios, and net present value) to ensure the survival and profitability of the firm and the project, and the survival of the engineering manager. This knowledge of basic financial and accounting principles and techniques not only enhances the company's communication processes, financial stability, and profitability, but acts as an incentive and motivation for engineers and engineering managers at all levels of management.

There must be cooperation and communication among all department project managers and middle and upper administration to ensure departmental and company objectives and goals. This responsibility can be assured by having all engineers with financial responsibilities be knowledgeable and involved in the financial process.

Jae K. Shim
Norman Henteleff

CONTENTS

WHAT EVERY ENGINEER SHOULD KNOW ABOUT

ACCOUNTING AND FINANCE

1

INTRODUCTION

Knowledge of basic accounting and finance will enable engineers and project managers to prepare, appraise, evaluate, and approve plans to accomplish departmental and company objectives. Most engineers who are promoted to management positions have years of training and experience in their technical areas but very limited orientations, education, or experience in cost analysis, budgeting, or project management. For the new engineering managers, the first item of business is survival. They must be knowledgeable and able to communicate effectively with accountants, financial officers, other engineers, and a host of upper and middle management people.

ACCOUNTING

Accounting is the language of the business. It is the process of recording, classifying, and analyzing the financial transactions of a business. Accounting is broadly divided into two main branches: financial accounting and managerial accounting.

Financial accounting is concerned mainly with the historical aspects of external reporting, that is, providing financial information to outside parties such as investors, creditors, and governments. To protect those outside parties from being misled, financial accounting is governed by what are called *generally accepted accounting principles (GAAP)*.

Management accounting, on the other hand, is concerned primarily with providing information to internal managers who are charged with planning and controlling the operations of the firm and making a variety of management decisions. Due to its internal use, management accounting is not subject to GAAP. More specifically, the differences between financial and management accounting are summarized below:

FINANCIAL ACCOUNTING	MANAGEMENT ACCOUNTING
1. Provides data for external users	1. Provides data for internal users
2. Is subject to GAAP	2. Is not subject to GAAP
3. Must generate accurate and timely data	3. Emphasizes relevance and flexibility of data
4. Looks backward	4. Looks forward
5. Looks at the business as a whole	5. Focuses on parts as well as on the whole of a business
6. Primarily stands by itself	6. Draws heavily from other disciplines such as finance, economics, and quantitative methods

Accounting and Finance

Finance or financial management is concerned with maximizing the wealth (or stock value) of the owner or stockholders of the business entity. More specifically, It covers areas that involve: (1) financial analysis and planning, (2) investment decisions, (3) financing and capital structure decisions, and (4) management of financial resources.

The financial manager of a firm plays an important role in the company's goals, policies, and financial success. The financial manager's responsibilities include:

1. Determining the proper amount of funds to employ in the firm, i.e., designating the size of the firm and its rate of growth
2. The efficient allocation of funds to specific assets
3. Raising funds on as favorable terms as possible, i.e., determining the composition of liabilities
4. Management of working capital such as inventory and accounts receivable.

Financial Officers

In a large firm, these financial responsibilities are carried out by the treasurer, controller, and financial vice president (chief financial officer). The treasurer is responsible for managing corporate assets and liabilities, planning the finances, budgeting capital, financing the business, formulating credit policy, and managing the investment portfolio. He or she basically handles external financing matters.

The controller is basically concerned with internal matters, namely, financial and cost accounting, taxes, budgeting, and control functions. The financial vice president, better known as a chief financial officer (CFO), supervises all phases of financial activity and serves as the financial adviser to the board of directors.

IMPORTANT ACCOUNTING AND FINANCE CONCEPTS

From the perspective of financial management there are some important concepts which include *return*, *risk*, *leverage*, and *cash flow*. The firm wants to be profitable and it wants to continue in business. Also, it may want to decide how much other people's money needs to be invested in the business to increase return to its owners. There exists, however, a tradeoff between return and the risk associated with a decision. Further, it is possible to be profitable and yet fail to continue in business because of lack of cash flow.

Risk vs. Return

In maximizing return, there is always a tradeoff with risk. The greater the risk to be taken, the greater the anticipated return we demand (Figure 1.1). For example, in the case of inventory management, the less inventory you keep, the higher the expected return (since less of the firm's current assets is tied up), but also the greater the risk of running out of stock and thus losing potential revenue.

Obviously, given two equally risky projects, we would choose the one with a greater expected return. The key in many business decisions is that you are faced with a tradeoff—risk vs. return. Decision makers, therefore, must ask, "Is the extra return worth the extra risk?"

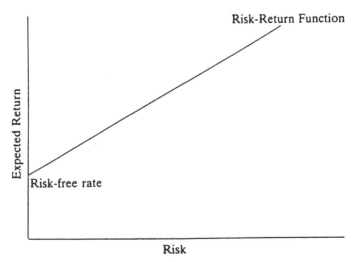

Figure 1.1 Return vs. risk.

Return vs. Liquidity

There also exists a tradeoff between liquidity and return. Greater liquidity results in more safety, but lower return.

As a means of increasing its liquidity, the business may choose to invest additional monies in cash and/or marketable securities such as T-bills and commercial papers. Such action involves a trade-off, since such assets are likely to earn little return.

Note that Company A has a current ratio of 1.73 and earns 13.16% return on its total assets. Company B, on the other hand, has a higher liquidity as expressed by a current ratio of 2.4, but earns only 12.09%.

Example 1.1.[*] Companies A and B are identical in every respect but one: Company B has invested $20,000 in T-bills, which has been financed with equity. Assume a 50% tax rate. The balance sheets and earnings of the two companies are shown in the A–B comparison chart.

	Company A	Company B
Cash	$ 2,000	$ 2,000
Marketable securities	—	20,000
Other current assets	50,000	50,000
Fixed assets	100,000	100,000
Total	$152,000	$172,000
Current liabilities	$ 30,000	$ 30,000
Long-term debt	50,000	50,000
Owner's equity	72,000	92,000
Total	$152,000	$172,000
Net income	$ 20,000	20,800 [a]
Current ratio	$\dfrac{\$\,52,000}{\$\,30,000} = 1.73$	$\dfrac{\$\,72,000}{\$\,30,000} = 2.4$
Return on total assets (ROA)	$\dfrac{\$\,20,000}{\$152,000} = 13.16\%$	$\dfrac{\$\,20,800}{\$172,000} = 12.09\%$

[a]During the year Company B held $20,000 in T-bills, which earned an 8% return or $1,600 for the year or $800 after taxes.

---

[*]This example may be omitted without loss of continuity. But readers are strongly advised to go back to this example after a thorough understanding of Chapters 2 and 4.

Leverage

One of the most important financial decisions faced by any business is how much leverage (use of other people's money) it should employ, i.e., the degree to which it incurs fixed costs. As sales increase, fixed costs do not increase. As a result, earnings can rise rapidly during good times. On the other hand, during bad times fixed costs do not decline, so profits fall sharply.

Leverage refers to the degree to which a firm commits itself to high levels of fixed costs. The more leverage a firm has, the riskier it is because of the obligations on fixed costs that must be met regardless of good times or bad times. By the same token, the more highly leveraged, the greater the profits during good times. Here again, there is a tradeoff between risk and return that the firm must consider in making a financial decision. There are two types of leverage, operating and financial, which will be discussed in detail in Chapter 7.

Cash vs. Accrual Basis of Accounting

To measure an entity's operating performance, it is necessary to match the revenue and the expense items applicable to a specified time period. There are two ways of recognizing revenue and expenses in financial reporting: cash basis and accrual basis. With a "cash basis", revenue is recorded only as cash is received, and expenses are recorded only as paid. With an accrual system, revenue is recognized and recorded when it is earned, even though it may not be billed or received for some time. Expenses are recorded when incurred, even though they may not be paid for some time.

Accountants generally recommend the accrual basis as giving the best indication of actual performance. Inherent in the accrual method of accounting is the so-called matching principle, which states that expenses should be matched against the revenue to which they are directly related, regardless of the timing of cash collections and payments. It poses a difficult problem, however, since the bottom line income figure does not represent cash flow. This problem is well illustrated in the following section.

You May Go Broke While Making a Profit!

If you are to manage cash flows, you must understand the difference between the profits (net income or earnings) you see on the bottom line of the income statement determined using the accrual basis of accounting and economic profits, and how accounting profits differ from economic profits, which are *cash flows*. The following example illustrates an important point you should know about: You can go broke while you show accounting profits!

Example 1.2.[*] As the year started, Mr. Geller of the Office Products Co. was in fine shape. His company made ball-point pens. He made them for $.75 each, sold them for $1. He kept a 30-day supply in inventory, *paid his bills promptly*, and billed his customers 30 days net. Sales were right on target, with the sales manager predicting a steady increase. It felt like his lucky year, and it began as shown in the balance sheet:

Office Products Company
Balance Sheet
January 1, 19x1

Cash	$1,000	Liabilities	0
Inventory	750		
Receivables	1,000	Retained earnings	$2,750
Total assets	$2,750	Total liabilities	
		and equity	$2,750

In January, he sold 1,000 ball-point pens, shipped them at a cost of $750 and collected his receivables—winding up with a tidy $250 profit and his books looked like this:

January 31, 19x1

Cash	$1,250	Liabilities	0
Inventory	750		
Receivables	1,000	Retained earnings	$3,000 ($2,750 + $250)
Total assets	$3,000	Total liabilities	
		and equity	$3,000

February's sales jumped, as predicted, to 1,500 ball-point pens. With a corresponding step-up in production to maintain his 30-day inventory, he made 2,000 pens at a cost of $1,500. All receivables from January were collected. The profit so far is: $625 ($250 + $375). Now his books looked like this:

February 28, 19x1

Cash	$ 750	Liabilities	0
Inventory	1,125		
Receivables	1,500	Retained earnings	$3,375 ($3,000 + $375)
Total assets	$3,375	Total liabilities	
		and equity	$3,375

[*]This example may be omitted without loss of continuity. But readers are strongly advised to go back to this example after a thorough understanding of Chapter 2.

March sales were even better: 2,000 units. Collections on time. Production, to adhere to his inventory policy: 2,500 units. Operating results for the month: $500 profit. Profit to date $1,125. His books now show:

March 31, 19x1

Cash	$ 375	Liabilities	0
Inventory	1,500		
Receivables	2,000	Retained earnings	$3,875 ($3,375 + $500)
Total assets	$3,875	Total liabilities	
		and equity	$3,875

In April, sales jumped another 500 units to 2,500 and Geller patted his sales manager on the back. His customers were paying right on time. Production was pushed to 3,000 units, and the month's business netted him $625 for a profit to date of $1,750. He took off to Florida before he saw the accountant's report:

April 30, 19x1

Cash	$ 125	Liabilities	0
Inventory	1,875		
Receivables	2,500	Retained earnings	$4,500 ($3,875 + $625)
Total assets	$4,500	Total liabilities	
		and equity	$4,500

May saw Geller' small business really hitting a stride—sales of 3,000 units, production of 3,500 and a five month profit of $2,500. But, suddenly, he got a phone call from his bookkeeper: Come home! We need money! His books had caught up with him:

May 31, 19x1

Cash	$ 0	Liabilities	0
Inventory	2,250		
Receivables	3,000	Retained earnings	$5,250 ($4,500 + $750)
Total assets	$5,250	Total liabilities	
		and equity	$5,250

To capture the critical interactions and relationships between net income and cash flow, presented below is the statement of cash flows which basically lists the sources and uses of cash. The message is clear: You can go broke while

OFFICE PRODUCTS COMPANY
Statement of Cash Flows
For the Month Ended June 30, 19x1

	Feb.	March	April	May	June	Total
Cash flows from operating activities:						
Net income(a)	$ 250	$ 375	$500	$625	$750	$2,500
Increase in:						
Inventory	0	500	500	500	500	2,000
Receivables	0	375	375	375	375	1,500
Total(b)	0	875	875	875	875	3,500
Incr.(Decr.)(c)	$ 250	($500)	($375)	($250)	($125)	($1,000)
Cash balance at the beginning(d)	1,000	1,250	750	375	125	1,000
Cash balance at the end(e)	$1,250	$ 750	$375	$125	$ 0	$ 0

Note: (c) = (a) − (b) and (e) = (c) + (d)
Source: Adapted from *Business Week* April 28, 1956, and G. Gallinger and P. Basil Healey, *Liquidity Analysis and Management*, Addison-Wesley, Reading, Mass., 1987.

making a profit. Geller's cash was down to zero, while the business made a five-month profit of $2,500.

BASIC FORMS OF BUSINESS ORGANIZATION

Accounting and finance are both applicable to all economic entities such as business firms and nonprofit organizations such as schools, governments, hospitals, churches, and so on. However, this book will focus on accounting and finance for business firms organized as three basic forms of business organizations. These forms are: (1) the sole proprietorship, (2) the partnership, and (3) the corporation.

Sole Proprietorship

This is a business owned by one individual. Of the three forms of business organizations, sole proprietorships are the greatest in number. The advantages of this form are:

1. No formal charter required
2. Less regulation and red tape

3. Significant tax savings
4. Minimal organizational costs
5. Profits and control not shared with others

The disadvantages are:

1. Limited ability to raise large sums of money
2. Unlimited liability for the owner
3. Limited to the life of the owner
4. No tax deductions for personal and employees' health, life, or disability insurance

Partnership

The partnership is similar to the sole proprietorship except that the business has more than one owner. Its advantages are:

1. Minimal organizational effort and costs
2. Fewer governmental regulations

Its disadvantages are:

1. Unlimited liability for the individual partners
2. Limited ability to raise large sums of money
3. Dissolution upon the death or withdrawal of any of the partners

There is a special form of partnership, called a *limited partnership*, where one or more partners, but not all, have limited liability up to their investment in the event of business failure.

1. The general partner manages the business.
2. Limited partners are not involved in daily activities. The return to limited partners is in the form of income and capital gains.
3. Often, tax benefits are involved.

Examples of limited partnerships are in real estate and oil and gas exploration.

Corporation

The corporation a legal entity that exists apart from its owners, better known as stockholders. Ownership is evidenced by possession of shares of stock. In terms of types of businesses, the corporate form is not the most common, but the most important in terms of total sales, assets, profits, and contribution to national income. Corporations are governed by a distinct set of state or federal laws and come in two forms: a state *C Corporation* or federal *Subchapter S.*

C Corporation

The advantages of a C corporation are:

1. Unlimited life
2. Limited liability for its owners, as long as no personal guarantee on a business-related obligation such as a bank loan or lease
3. Ease of transfer of ownership through transfer of stock
4. Ability to raise large sums of capital

Its disadvantages are:

1. Difficult and costly to establish, as a formal charter is required
2. Subject to double taxation on its earnings and dividends paid to stockholders
3. Bankruptcy, even at the corporate level, does not discharge tax obligations

Subchapter S Corporation

It is a form of corporation whose stockholders are taxed as partners. To qualify as an S corporation, the following is necessary:

1. A corporation cannot have more than thirty-five shareholders.
2. It cannot have any nonresident foreigners as shareholders.
3. It cannot have more than one class of stock.
4. It must properly elect Subchapter S status.

The S corporation can distribute its income directly to shareholders and avoid the corporate income tax while enjoying the other advantages of the corporate form. *Note*: Not all states recognize Subchapter S corporations.

TOPICS COVERED IN THIS BOOK

This book covers the following topics:

Chapter 2. Understanding Financial Statements

The engineer should have a good understanding of the company in order to make an informed judgement on the financial position and operating performance of the entity. The balance sheet, the income statement, and the statement of cash flows are the primary documents analyzed to determine the company's financial condition. The balance sheets gives the company's position in terms of its assets, liabilities, and equity or net worth, while the income statement gives the company's sources of revenue, expenses, and net income. The statement of cash flows allows you to analyze the company's sources and uses of cash. These financial statements are included in the annual report.

The allocation of engineering and research costs require the combined efforts of the technical department and the accounting department. The basic accounting formula for a profit-oriented enterprise is simply: Profit = Revenue − Expenses. Revenue comes primarily from sales of goods and services (operating income) and miscellaneous income such as interest and sales of items the company has used but no longer needs. Expenses are usually shown on the income statement as (1) cost of goods sold, (2) selling costs and, (3) administrative costs. Engineering costs are often buried in the latter, except when they apply to specific products, then they are included in the cost of goods sold. Therefore, it is advisable for all engineers who have responsibility for expenditure of funds to have a basic knowledge of cost elements to be able to understand and prepare cost estimates and reports.

Chapter 3. Recording Financial Information and Accounting Conventions

Financial decisions are usually formulated on the basis of information generated by the accounting system of the firm. Proper interpretation of the data requires an understanding of the assumptions and rules underlying such systems, the convention adopted in recording information, and the limitation inherent in the information presented. To facilitate this understanding, basic accounting concepts and conventions are presented and discussed in this chapter.

Chapter 4. Analyzing Financial Statements

You also need to be able to analyze your company's financial statements in order to evaluate its financial health and operating performance. What has been the trend in profitability and return on investment? Will the business be able to pay its bills? How are the receivables and the inventory turning over? This chapter presents various financial statement analysis tools that are useful in evaluating the company's current and future financial conditions. These techniques include horizontal, vertical, and ratio analysis.

Chapter 5. Cost Concepts, Cost Behavior, and Cost Accounting

You need to realize that there are different costs used for different purposes. Some costs are useful and required for inventory valuation and income determination. Some costs are useful for planning, budgeting, and cost control (e.g., variable costs, fixed costs, and standard costs). Still others are useful for making decisions (e.g., opportunity costs and sunk costs). You need to be familiar with (1) various cost terms in order to make informed decisions about cost control and (2) how manufacturing costs are accumulated in order to interpret the cost data properly. And then cost analysis can be applied to appraising and evaluating costs associated with a decision or project including evaluating effects upon profitability. The chapter addresses these issues.

Chapter 6. Budgeting for Profit Planning

You must be able to express your budgetary needs in order to obtain proper funding for your department. You may have to forecast future sales, cash flows, and costs, to see if you will be operating effectively in the future. Budgets are closely tied to technical task descriptions—they represent the economic and time qualification of the technical plan. From the view point of the engineer or engineering manager, the budget offers a means to do a better job and prevent failure. The reporting of performance against the budget goals and standards permits measurement and control of engineering activities. Budgeting and cost reporting lead to better definitions of objectives, more effective evaluation of work in progress and better measurement of performance at every level.

Budgets begin with planning since planning is the key to good budgeting and cost control. Each project, product or service must be estimated in terms of costs or dollar amounts. The budget helps in allocating the funds and selecting from among alternative projects. It identifies financial differences between actuality and plans. It establishes controls that prevent expenditures not in accordance with the approved plans. It allows for prospective changes. It helps identify most engineering management problems which tend to be dollar problems since they are tied to financial considerations. This chapter explains how to develop a budget, step by step, and how it can be used as a control device.

Chapter 7. Cost-Volume-Profit Analysis and Leverage

One of the most important aspects in business is the break-even concept. Break-even analysis is a method for analyzing the relationship between fixed costs, variable costs, sales volume, and profit. It is often used in decisions such as the choice between two different production processes, or types of machines, the choice between producing an item internally or purchasing it from a supplier, determining whether to buy or lease, determining the size or number of facilities to be incorporated in the manufacturing system, or seeing if a project or program breaks even. In this chapter, you will learn the mechanics of break-even analysis and how it can help you make various manufacturing-related choices.

Chapter 8. Responsibility Accounting and Cost Control Through Standard Costs

A comparison of actual costs to standard (budgeted) costs is helpful in identifying problem areas. Variance analysis helps you to spot areas of inefficiency by comparing actual performance to standards. What are the cost overruns? Are they significant enough to warrant your investigation? Are they controllable? Who is responsible? What are the reasons for these cost overruns? Variances of any kind may be related directly to technical areas. The cost should be segregated into direct costs and indirect costs (or overhead or burden). Direct costs consist primarily of the salaries of technical people working on the particular job plus the materials used for the job (direct materials). It is particularly important that cost reporting in technical organizations be in terms of direct costs. Accounts or "shop orders" identified by titles and code numbers are established to permit collections of costs at the smallest subdivision of element costs desired. Cost reporting is an important phase of measuring and evaluating engineering projects. This chapter discusses the mechanics of calculating variances, how to analyze them for performance improvement, and the concept of flexible budgeting.

Chapter 9. Improving Divisional Performance

The engineer and engineering manager should have a basic understanding of financial information so as to evaluate the performance of his area of responsibility. Are things getting better or worse? What are the possible reasons? Who is responsible? What can be done about it? Certain strategies can be undertaken to improve the return on investment of your department by enhancing profitability or using your assets more efficiently. This chapter presents two alternative measures of divisional performance, return on investment (ROI) and residual income (RI). It concentrates on various strategies for improving divisional performance. Also covered is the issue of transfer pricing which may significantly affect divisional performance.

Chapter 10. Relevant Costing in Nonroutine Decisions

In this chapter, you, as a project manager or engineer, will learn how to handle many common production choices you may face. These non-routine decisions include such actions as whether to make or buy, to sell a product at below the normal selling price, to keep or drop a certain production line, or whether to sell a product at the split-off point or process further. Two key tools for cost analysis are relevant costs and contribution analysis, which are discussed in this chapter. Also presented is a discussion of how to utilize limited resources such as productive capacity, warehouse space, or skilled labor.

Chapter 11. Applying the Time Value of Money

As a project manager you will need to understand and recognize the time value of money—the idea that a dollar is worth less the longer it takes to receive it, which sets the basis for a variety of financial decisions including capital expenditure decisions. Thus, you would prefer projects that generate higher cash flows in earlier years. Time value of money applications include determining the present value of receiving future cash flows, computing how much money will be in an account at a future date, and calculating interest rates, periodic payments on a loan, and the time it will take for money to grow to a specific value.

Chapter 12. Evaluating Capital Investment Projects

You are often faced with a choice of facilities as investment opportunities. Decisions to invest in new facilities may arise from several sources. First, you may need to expand the company's productive capacity. Second, you may need additional technical capabilities in the form of new manufacturing processes, research facilities, computerized information systems, and the like. Third, you may need to replace inefficient equipment with high-tech equipment. Fourth, you may have to decide whether to buy machine A or machine B, whether to introduce a certain product line, or whether to buy or lease. You, as a project engineer or manager, will have to consider not only the time value of money, but also the role of depreciation and income taxes in investment decisions. This chapter examines the process of evaluating facilities investment projects, including the use of such techniques as the uniform annual cost, and discounted cash flow (DCF) techniques. The problems that arise with mutually exclusive investments and income tax factors are also addressed. Also, you must have an understanding of how to determine the cost of capital which is used as the discount rate in the DCF and residual income (RI) calculations.

Chapter 13. How Taxes Affect Business Decisions

Taxes are important in any business decision; the after-tax effect is what counts. Proper tax planning will facilitate wise decisions. Are you maximizing your allowable tax deductions? This chapter discusses sources of tax-exempt income, tax-deductible expenses, timing income and expenses among tax years, and some tax planning strategies. The chapter emphasizes the new Clinton tax based on the *1993 Revenue Reconciliation Act.*

2

UNDERSTANDING THE FINANCIAL STATEMENTS

Typically, most businesses rely on two financial statements—the income statement and the balance sheet—to tell them how well they have done and where they stand. An examination of what can be gained from these statements, and where the pitfalls lie, is useful in setting up a program or strategy for planning and controlling profits.

THE INCOME STATEMENT AND BALANCE SHEET

The income statement measures operating performance for a specified time period (like for the year ended December 31, 19x1). The income statement shows the revenue, expenses, and net income (or loss) for a defined period for each of the following elements.

Revenue

Revenue arises from the sale of merchandise (as by retail business), or the performance of services for a customer or a client (as by a lawyer). Revenue from sales of merchandise or sales of services is often identified merely as *sales*. Other terms used to identify sources of revenue include professional fees, commission revenue, and fares earned. When revenue is earned it results in an increase in either Cash or Accounts Receivable.

Expenses

Expenses result from performing those functions necessary to generate revenue. The amount of an expense is either equal to the cost of goods sold, the value of the services received (e.g., salary expense), or the expenditures necessary for conducting business operations (e.g., rent expense), during the period.

15

Net Income (Loss)

Net Income, also called *profit* or *earnings*, is the amount by which total revenue exceeds total expenses for the reporting period. It should be noted that revenue does not necessary mean receipt of cash and expense does not automatically imply a cash payment. Note that net income and net cash flow (cash receipts less cash payments) are different. For example, taking out a bank loan will generate cash but this is not revenue since merchandise has not been sold nor have services been provided. Furthermore, capital has not been altered because of the loan.

Note that each revenue and expense item has its own account. This specifically enables one to better evaluate and control revenue and expense sources and to examine relationships among account categories. For instance, the ratio of telephone expenses to revenue is 10% ($1,000/$10,000). If in the previous month the relationship was 3%, Joan Biehl would, no doubt, attempt to determine the cause for this significant increase.

The balance sheet, on the other hand, portrays the financial position of the company at a particular point in time. It shows what is owed (assets), how much is owed (liabilities), and what is left (assets minus liabilities, known as stockholders' equity or net worth). With the balance sheet, you cut the point, freeze the action, and want to know about the company's financial position as of a certain date (like 12/31/19x1, the end of the reporting year). It is a snapshot, while the income statement is a motion picture.

Example 2.1. Joan Biehl is a self-employed engineer. For the month of May 19X1, she earned income of $10,000 from services rendered. Her business expenses were: telephone $1,000, electricity $500, rent $2,000, secretarial salary $300, and office supplies used, $400. Her income statement for the period is shown in the income statement:

Joan Biehl

Income Statement

For the Month Ended May 31, 19x1

Revenue from professional services		$10,000
Less: Operating expenses		
Telephone	$ 1,000	
Electricity	500	
Rent	2,000	
Secretarial salary	300	
Office supplies	400	
Total operating expenses		4,200
Net income		$ 5,800

Assets

Assets are economic resources which are owned by an organization and are expected to benefit future operations. Assets may have definite physical form such as buildings, machinery, or supplies. On the other hand, some assets exist not in physical or tangible form, but in the form of valuable legal claims or rights, such as *accounts receivables* from customers and *notes receivables* from debtors.

Assets which will be converted into cash within one year are classified as *current*. Examples are cash, marketable securities, receivables, inventory, and prepaid expenses. Prepaid expenses include supplies on hand and advance payments of expenses such as insurance and property taxes.

Assets having a life exceeding one year are classified as *noncurrent*. Examples are long-term investments, equipment, and buildings. Equipment and buildings are often called *plant assets* or *fixed assets*.

Liabilities

Liabilities are debts, owed to outsiders (creditors) and are frequently described on the balance sheet by titles that include the word "payable." The liability arising from the purchase of goods or services on credit (on "time") is called an *accounts payable*. The form of liability when money is borrowed is usually a *note payable*, a formal written promise to pay a certain amount of money, plus interest, at a definite future time. *Accounts payable*, as contrasted with a *note payable*, does not involve the issuance of a formal promise written to the creditor, and it does not require payment of interest. Other examples of liabilities include various accrued expenses.

Liabilities payable within one year are classified as *current*, such as accounts payable, notes payable, and taxes payable. Obligations payable in a period longer than one year, for example, bonds payable, are termed *long-term liabilities*.

Equity

Equity is a residual claim against the assets of the business after the total liabilities are deducted. Capital is the term applied to the owner's equity in the business. Other commonly used terms for capital are *owner's equity* and *net worth*. In a sole proprietorship, there is only one capital account since there is only one owner. In a partnership, a capital account exists for each owner. In a corporation, capital represents the *stockholders' equity*, which equals the capital stock issued plus the accumulated earnings of the business (called retained earnings). There are two types of capital stock—common stock and preferred stock. Common stock entitles its owners to voting rights, while preferred stock does not. Preferred stock entitles its owners to priority in the receipt of dividends and in repayment of capital in the event of corporate dissolution.

The balance sheet may be prepared either in *report form* or *account form*. In the report form, assets, liabilities, and capital are listed vertically. In the account form, assets are listed on the left side and liabilities and capital on the right side.

From the examples given, it is evident that there is a tie-in between the income statement and the balance sheet. Biehl's net income of $5,800 (the last item in her income statement from Example 2.1) is added to capital in her balance sheet in the above example. In effect, the income statement serves as the bridge between two consecutive balance sheets. Further, the net balance of the income statement accounts is used to adjust the Capital Account in the Journal accounts (Chapter 3).

Example 2.2. The equity of the owners of the business is quite similar to the equity commonly referred to with respect to home ownership. If you were to buy a house for $150,000 by putting down 20%, i.e., $30,000 of your own money and borrowing $120,000 from a bank, you would say that your equity in the $150,000 house was $30,000.

Report Form

Joan Biehl Balance Sheet May 31, 19x1			
Assets			
Cash			$10,000
Accounts receivable			20,000
Office supplies			10,500
Office equipment			30,000
Total assets			$71,000
Liabilities and Capital			
Liabilities			
Accounts payable			$30,000
Capital			
Balance, May 1, 19x1		$35,600	
Net income for May	$5,800		
Less withdrawals	400		
Increase in capital	$5,400		
Total capital			41,000
Total liabilities and capital			$71,000

Account Form

Joan Biehl
Balance Sheet
May 31, 19x1

Assets   Liabilities and Capital

Cash	$10,000	Liabilities		
Accounts receivable	20,000	Accounts payable		$30,000
Office supplies	10,500	Capital		
Office equipment	30,000	Balance, May 1,19x1		$35,600
		Net income for May:	$5,800	
		Less: Withdrawals	400	
		Increase in capital		5,400
		Total capital		41,000
Total Assets	$71,000	Total Liabilities & Capital		$71,000

Figure 2.1 shows the relationship between the income statement and the balance sheet. In fact, the income statement serves as a bridge between the two consecutive balance sheets. *Note*: Simply put, with the balance sheet you are asking "how wealthy or poor is the company?," while with the income statement you are asking "how did the company do last year?" and "did it make money and then how much?" Neither one is good enough to tell you about the financial health of the company. For example, the fact that the company made a big profit does not necessarily mean it is wealthy, and vice versa. In order to get the total picture, you need both statements to complement each other.

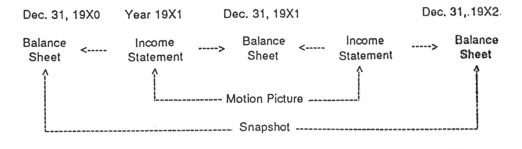

Figure 2.1 Balance Sheet and Income Statement Relationship.

Unfortunately, you still have problems. You would like to know more about the company's financial shape (such as the cash position of the company). However, neither the balance sheet nor the income statement provides information about the flow of cash during the period. The statement of cash flows provides this information, which will be discussed below.

THE STATEMENT OF CASH FLOWS

The statement of cash flows shows the sources and uses of cash, which is a basis for cash flow analysis for managers. The statement aids you in answering vital questions such as "where was money obtained?" and "where was money put and for what purpose?" The following provides a list of more specific questions that can be answered by the statement of cash flows and cash flow analysis:

1. Is the company growing or just maintaining its competitive position?
2. Will the company be able to meet its financial obligations?
3. Where did the company obtain funds?
4. What use was made of net income?
5. How much of the required capital has been generated internally?
6. How was the expansion in plant and equipment financed?
7. Is the business expanding faster than it can generate funds?
8. Is the company's dividend policy in balance with its operating policy?
9. Is the company's cash position sound and what effect will it have on the market price of stock?

Cash is vital to the operation of every business. How management utilizes the flow of cash can determine a firm's success or failure. Financial managers must control their company's cash flow so that bills can be paid on time and extra dollars can be put into the purchase of inventory and new equipment or invested to generate additional earnings.

FASB Requirements

Management and external interested parties have always recognized the need for a cash flow statement. Therefore, in recognition of the fact that cash flow information is an integral part of both investment and credit decisions, the Financial Accounting Standards Board (FASB) has issued Statement No. 95, "Statement of Cash Flows." This pronouncement requires that enterprises include a statement of cash flows as part of their financial statements. A statement of cash flows reports the cash receipts, payments, and net change in cash on hand resulting from the *operating*, *investing*, and *financing* activities of an enterprise during a given period. The presentation reconciles beginning and ending cash balances.

Accrual Basis of Accounting

Under Generally Accepted Accounting Principles (GAAP), most companies use the accrual basis of accounting. This method requires that revenue be recorded when earned and that expenses be recorded when incurred. Revenue may include credit sales that have not yet been collected in cash and expenses incurred that may not have been paid in cash. Thus, under the accrual basis of accounting, net income will generally not indicate the net cash flow from operating activities. To arrive at net cash flow from operating activities, it is necessary to report revenues and expenses on a cash basis. This is accomplished by eliminating those transactions that did not result in a corresponding increase or decrease in cash on hand.

Example 2.3. During 19x1, the Eastern Electric Supply Corporation earned $2,100,000 in credit sales, of which $100,000 remained uncollected as of the end of the calendar year. Cash that was actually collected by the corporation in 19x1 can be calculated as follows:

Credit sales	$2,100,000
Less: Credit sales uncollected at year end	100,000
Actual cash collected	$2,000,000

A statement of cash flows focuses only on transactions involving the cash receipts and disbursements of a company. As previously stated, the statement of cash flows classifies cash receipts and cash payments into operating, investing, and financing activities.

Operating Activities

Operating activities include all transactions that are not investing or financing activities. They only relate to income statement items. Thus cash received from the sale of goods or services, including the collection or sale of trade accounts and notes receivable from customers, interest received on loans, and dividend income are to be treated as cash from operating activities. Cash paid to acquire materials for the manufacture of goods for resale, rental payments to landlords, payments to employees as compensation, and interest paid to creditors are classified as cash outflows for operating activities.

Investing Activities

Investing activities include cash inflows from the sale of property, plant, and equipment used in the production of goods and services, debt instruments or equity of other entities, and the collection of principal on loans made to other enterprises. Cash outflows under this category may result from the purchase of plant and equipment and other productive assets, debt instruments or equity of other entities, and the making of loans to other enterprises.

Financing Activities

The financing activities of an enterprise involve the sale of a company's own preferred and common stock, bonds, mortgages, notes, and other short- or long-term borrowings. Cash outflows classified as financing activities include the repayment of short- and long-term debt, the reacquisition of treasury stock, and the payment of cash dividends.

Example 2.4. The following information pertains to the Liverpool Sugar Corporation during 19x1.

1. The company had $1,004,000 in cash receipts from the sale of goods. Cash payments to acquire materials for the manufacture of goods totaled $469,000, its payments on accounts and notes payable amounted to $12,000, and it paid $136,000 in federal and state taxes.
2. The company sold all of its stock investment in Redondo Food Corporation, an unrelated entity, for $100,000. It then bought a new plant and equipment for $676,000.
3. In 19x1, the company sold $300,000 of its 10%, ten-year bonds. It also issued another $50,000,000 in preferred stock in return for land and buildings. The company paid a cash dividend of $36,000.
4. The company had a $198,000 cash balance at the beginning of the year.

The statement cash flows for the company would be presented as shown in the display statement. Note that the issuance of the preferred stock in exchange for the land and buildings is a noncash transaction that would be disclosed in supplementary form at the end of the statement separate from cash flows.

Liverpool Sugar Corporation
Statement of Cash Flows
for the Year Ended December 31, 19x1

Cash flows from operating activities:		
Cash received from customers	$ 1,004,000	
Cash payment for acquisition of materials	(469,000)	
Cash payment for interest and dividends	(12,000)	
Cash payment for taxes	(136,000)	
Net cash provided by operating activities		$ 387,000

Cash flows from investing activities:		
Cash paid to purchase plant and equipment	$(676,000)	
Sale of long-term investment	(100,000)	
Net cash provided by investing activities		(776,000)
Cash flows from financing activities:		
Sale of bonds	$ 300,000	
Cash paid for dividends	(36,000)	
Net cash used in financing activities		264,000
Net decrease in cash and cash equivalents		$(125,000)
Cash and cash equivalents at the beginning of the year.		198,000
Cash and cash equivalents at the end of the year		$ 73,000

NOTES TO FINANCIAL STATEMENTS

Accounting numbers do not always tell the entire story. For a variety of reasons these three financial statements reported in an annual report tend to be inadequate to fully convey the results of operations and the financial position of the firm.

The annual report often contains this statement: "The accompanying footnotes are an integral part of the financial statements." This is because the financial statements themselves are concise and condensed. Hence, any explanatory information that cannot be readily abbreviated is provided for in greater detail in the footnotes.

Footnotes provide detailed information regarding financial statement figures, accounting policies, explanatory data such as mergers and stock options, and furnish any additional necessary disclosure. Examples of footnote disclosures are accounting methods and estimates such as inventory pricing, pension fund and profit sharing arrangements, terms of characteristics of long-term debt, particulars of lease agreements, contingencies, and tax matters.

The footnotes appear at the end of the financial statements and *explain* the figures in those statements. Footnote information may be in both *quantitative* and *qualitative* terms. An example of quantitative information is the fair market value of pension plan assets. An example of a qualitative disclosure is a lawsuit against the company. It is essential that the reader carefully evaluate footnote information to derive an informed opinion about the company's financial health and operating performance.

CONCLUSION

The traditional accounting statements—balance sheet, income statement and the newly required statement of cash flows—have been and will continue to be the most important tools for both management and outsiders for use in gauging the financial condition of a business. As a later chapter will show, additional insights in the performance of the business can be gained by using ratio analysis. Importance in cash flow cannot be emphasized enough and much of the chapter was devoted to this topic.

In viewing the various figures on financial statements, the footnotes to a statement must be considered an integral part of the statement. Failure to consider the additional facts set forth by footnotes may lead to erroneous conclusions.

3

RECORDING FINANCIAL INFORMATION AND ACCOUNTING CONVENTIONS

The transactions of most businesses are numerous and complex, affecting many different items appearing on the financial statements. Therefore, a formal system of classification and recording is required for timely financial reporting and managerial needs. The aim of this chapter is to introduce the formal classification system of financial information commonly called *double-entry accounting*. By acquiring background information about this system, you will be able to more clearly understand the basic structure of the financial statements which were discussed in the previous chapter. In doing so, we will discuss some basic accounting conventions and also touch on popular accounting terms and concepts.

DOUBLE-ENTRY AND THE ACCOUNTING EQUATION

Double-entry accounting is a system in which each business transaction affects and is recorded in two or more accounts with equal debits and credits.

The Accounting Equation

An entity's financial position is reflected by the relationship between its assets and its liabilities and equity.

The accounting equation reflects these elements by expressing the equality of assets to creditors' claims and owners' equity as follows:

Assets = Liabilities + Equity

The equation in effect says that a company's assets are subject to the rights of debt holders and owners.

The accounting equation is the basis for double entry accounting, which means that each transaction has a duel effect. A transaction affects either both sides of the equation by the same amount or one side of the equation only, by increasing and decreasing it by identical amounts and thus netting zero.

Example 3.1. Foster Architectural Company has assets of $700,000, and obligations of $300,000, and owner's equity of $400,000. The accounting equation is

Assets = Liabilities + Equity
$700,000 = $300,000 + $400,000

If at the end of the reporting period, the firm derived net income of $80,000, the accounting equation becomes

Assets = Liabilities + Equity
$780,000 = $300,000 + $480,000

If $60,000 was then used to pay creditors, the accounting equation becomes

Assets = Liabilities + Equity
$720,000 = $240,000 + $480,000

In the next example, we will illustrate how the transactions of a business are recorded and what effect they have on the accounting equation.

Example 3.2. Lloyd Thomas, a consulting engineer, experienced the following events in the month of January 19x1:

1. Started his engineering consulting practice with a cash investment of $10,000 and office equipment worth $5,000
2. Purchased office supplies of $800 by paying cash
3. Bought a word processor for $500 on account from Smith Corona
4. Paid $400 in salary to his staff
5. Received an electric bill for $300
6. Earned professional fees of $20,000, of which $12,000 was owed
7. Paid $300 to Smith Corona
8. Withdrew $100 from the firm for personal use
9. Received $1,000 from one of the clients who owed him money

The transactions will now be analyzed.

Transaction 1. Thomas started his engineer consulting practice by investing $10,000 in cash and $5,000 in office equipment. The *assets* Cash and Office

Equipment are increased, and the *equity* is also increased for the total investment of the owner.

Assets (A)		=	Liabilities (L)	+	Equity (E)
Cash	Office Equipment (OE)				L. Thomas Equity (E)
$10,000	$5,000				$15,000

Transaction 2. Acquired office supplies for cash, $800.

The asset Office Supplies goes up by $800 with a corresponding reduction in the asset Cash. This is an example of one asset being used to acquire another one.

A			=	L	+	E
Cash	OE	Office Supplies (OS)				
$10,000	$5,000					$15,000
−800		$800				
$ 9,200	$5,000	$800				$15,000

Transaction 3. Purchased a word processor on account, $500.

An asset, Office Equipment, is being acquired on credit, thereby creating a *liability* for future payment called Accounts Payable. Accounts Payable is defined as the amount owed to suppliers.

A			=	L	+	E
Cash	OE	OS		Accounts Payable (AP)		
$9,200	$5,000	$800				$15,000
	+500			$500		
$ 9,200	$5,500	$800		$500		$15,000

Transaction 4. Paid salary, $400.

Cash and equity are both being reduced because of the wage expense. Equity is reduced because expenses of the business decrease the equity of the owner.

	A		=	L	+	E
Cash	OE	OS		AP		
$9,200	$5,500	$800		$500		$15,000
−400	+500					−400
$ 8,800	$5,500	$800		$500		$14,600

Transaction 5. Received an electric bill for $300 (not paid).

Liabilities are increased by $300 since the firm *owes* the utility money for electricity supplied. Equity is reduced for the *expense*.

	A		=	L	+	E
Cash	OE	OS		AP		
$8,800	$5,500	$800		$500		$14,600
				+300		−300
$ 8,800	$5,500	$800		$800		$14,300

Transaction 6. Earned fees of $20,000, of which $12,000 was received in cash and $8,000 was owed by clients.

Cash goes up by $12,000 and the Accounts Receivable (amounts owed to the business from customers) is created. Professional fees earned is *revenue* to the business and hence increases the owner's equity. Thus, equity is increased by $20,000.

	A			=	L	+	E
Cash	OE	OS	AR		AP		
$8,800	$5,500	$800			$800		$14,300
+12,000			$8,000				+20,000
$ 20,800	$5,500	$800	$8,000		$800		$34,300

Transaction 7. Paid $300 to Smith Corona (in partial payment of the amount owed to them).

The payment lowers the asset Cash and reduces the liability Accounts Payable.

	A			=	L	+	E
Cash	OE	OS	AR		AP		
$20,800	$5,500	$800	$8,000		$800		$34,300
−300					−300		
$ 20,800	$5,500	$800	$8,000		$500		$34,300

Transaction 8. Withdrew $100 for personal use.

Cash is reduced and so is equity. The personal withdrawal is, in effect, a disinvestment in the business and hence reduces equity. It is not an expense in running the business.

A				=	L	+	E
Cash	OE	OS	AR		AP		
$20,500	$5,500	$800	$8,000		$500		$34,300
−100							−100
$ 20,400	$5,500	$800	$8,000		$500		$34,200

Transaction 9. Received $1,000 from clients who owed him money.

This increases Cash and reduces Accounts Receivable since the client now owes the business less money. One asset is being substituted for another one.

A				=	L	+	E
Cash	OE	OS	AR		AP		
$20,400	$5,500	$800	$8,000		$500		$34,200
+1,000			−1,000				
$ 21,400	$5,500	$800	$7,000		$500		$34,200

Transaction 10. Worth (determined by an inventory count) of office supplies on hand at month's end, $600.

Since the worth of office supplies originally acquired was $800 and $600 is left on hand, the business used $200 in supplies. This reduces the asset Office Supplies and correspondingly reduces equity. The supplies used up represent office supplies expense.

A				=	L	+	E
Cash	OE	OS	AR		AP		
$21,400	$5,500	$800	$7,000		$500		$34,200
		−200					−200
$ 21,400	$5,500	$600	$7,000		$500		$34,000

Lloyd Thomas
Summary of Transactions
January 19x1

	A			=	L	+ E
Cash	OE	OS	AR		AP	
1. $10,000	$5,000					$15,000
2. −800		$800				
$ 9,200	$5,000	$800				$15,000
3.	+500				$500	
$ 9,200	$5,500	$800			$500	$15,000
4. −400						−400 Wage Expense
$ 8,800	$5,500	$800			$500	$14,600
5.					+300	−300 Utility Expense
$ 8,800	$5,500	$800			$800	$14,300
6. +12,000			$8,000			+20,000 Prof. Fee, Revenue
$20,800	$5,500	$800	$8,000		$800	$34,300
7. −300					−300	
$20,500	$5,500	$800	$8,000		$500	$34,300
8. −100						−100 Drawing
$20,400	$5,500	$800	$8,000		$500	$34,200
9. +1,000			−1,000			
$21,400	$5,500	$800	$7,000		$500	$34,200
10.		−200				−200 Office Supplies Expense
$21,400	$5,500	$600	$7,000	=	$500	$34,000

$34,500		=		$34,500

The Account

To prepare the equation Assets = Liabilities + Stockholders' Equity for each transaction would be extremely time consuming. In addition, information about a specific item (e.g., accounts receivable) would be lost through this process. Instead, there should be an *account* established for each type of item. At the end of the reporting period, the financial statements can then be prepared based upon the balances in these accounts.

The basic component of the formal accounting system is the *account*. A separate account exists for each item shown on the financial statements. Thus, balance sheet accounts consists of assets, liabilities, and equity. Income statement accounts are either expenses or revenue. The increases, decreases, and balance are shown for each account.

In other words, the purpose of the account is to provide a capsule summary of all transactions which have caused an increase or decrease and to reflect the account balance at any given point in time.

Ledger

All accounts are maintained in a book called the *ledger*. The ledger of a firm, for example, would be the group of accounts which summarize the financial operations of the company and is the basis for the preparation of the balance sheet and income statement. It is also useful for decision making since it provides the manager with the balance in a given account at a particular time.

A Chart of Accounts

The ledger is usually accompanied by a table of contents called a *chart of accounts*. The chart of accounts is a listing of the titles and numbers of all accounts in the ledger. Listed first are the balance sheet accounts—assets, liabilities, and stockholders' equity, in that order. The income statement accounts—revenue and expenses—follow. Its numbering system permits easy reference to accounts.

The account numbering system as it pertains to a typical company is as follows:

Series	Account classification
001-199	Assets
200-299	Liabilities
300-399	Revenue
400-499	Expenses
500-599	Stockholders' equities

Particular accounts can then be given unique, identifying account numbers within a series, as in the following examples from the Assets Series:

Account no.	Assets
001	Cash on hand
002	Marketable securities
003	Accounts receivable
004	Inventories
005	Investments
006	Land
007	Buildings
008	Equipment

Accounts may take many forms, but the simplest is called a *T-account*. The reason for this name is obvious as shown below:

Account title

Debit (left side)	Credit (right side)

Every account has three major parts:

1. A title, which is the name of the item recorded in the account
2. A space for recording increases (monetary) in the amount of the item
3. A space for recording decreases (monetary) in the amount of the item.

The left and right sides of the account are called *debit* and *credit*, respectively (often abbreviated as "Dr" for debit and "Cr" for credit). Amounts entered on the left side of an account, regardless of the account title, are called *debits* to the account, and the account is said to be *debited*. Amounts entered on the right side of an account are called *credits*, and the account is said to be *credited*. You must note that the items debit and credit are *not* synonymous with the words increase and decrease. The system of debits and credits as related to increases and decreases in each of the five categories of accounts—assets, liabilities, revenue, expenses, and equity—and is explained later in the chapter.

To illustrate the account, we'll look at the Cash account (within the Asset classification), where receipts of cash during a period of time have been listed vertically on the debit side and the cash payments for the same period have been listed similarly on the credit side of the account. A memorandum total of the cash receipts for the period to date, $55,000 in the illustration, may be noted below the last debit whenever the information is desired.

Cash

	10,000	25,000
	45,000	21,000
		7,500
	55,000	53,500
1,500		

The total of the cash payments, $53,500 in the illustration, may be noted on the credit side in a similar manner. Subtraction of the smaller sum from the larger, $55,000 - $53,500, yields the amount of cash on hand, which is called the balance of the account. The cash account in the illustration has a *balance* of $1,500,(which may be inserted as shown),which identifies it as a *debit balance*.

The System of Debits and Credits

In this section we will explain how accounts are increased or decreased through the use of debits and credits, the basic foundation of *double-entry* accounting where at least *two* entries, a debit and a credit, are made for each transaction.

The following guide shows how to increase or decrease accounts using debits and credits.

Asset		Liabilities	
+	−	+	−
Debit for increase	Credit for decrease	Debit for decrease	Credit for increase

Revenue		Expenses	
+	−	+	−
Debit for decrease	Credit for increase	Debit for increase	Credit for decrease

Equity	
+	−
Debit for decrease	Credit for increase

These same relationships are illustrated on the following page:

Type of account	Normal balance	To increase	To decrease
Asset	Debit	Debit	Credit
Liability	Credit	Credit	Debit
Revenue	Credit	Credit	Debit
Expenses	Debit	Debit	Credit
Equity	Credit	Credit	Debit

The illustrated system of debits and credits is the standard method followed by persons keeping records on the double-entry system. The system of rules is analogous to a set of traffic rules whereby everyone (at lease everyone in this country) agrees to drive on the right side of the road. Obviously, the system would work if we reversed everything. However, you will see shortly that there is a very logical and unique system in the present structure.

The "How and Why" of Debits and Credits

Review the fundamental accounting equation:

Assets (A) = Liabilities (L) + Equity (E)

In addition to this equation there is another fundamental accounting concept or rule:

Debits must always equal Credits.

This means that whenever a financial transaction is recorded in the accounting record one account (or accounts) must be debited and another account (or account) must be credit to obtain an equal amount. It was noted earlier that all accounts have two sides, a debit side and a credit side. The purpose is to record increases on one side and decreases on the other.

Illustration of Debit and Credit Analysis

The following illustration of debit and credit analysis uses the transactions given in Example 3.2 for Lloyd Thomas, a consulting engineer. Each transaction is stated, analyzed, and followed by an illustration of the appropriate debit and credit (journal) entries in the various accounts, using T-accounts for simplicity. Each transaction has been numbered for reference as in Example 3.2. In the transaction analysis and the resulting debits and credits, each entry resulting from a particular transaction is parenthetically keyed to the transaction number.

Transaction 1. Thomas started his engineer consulting practice by investing $10,000 in cash and $5,000 in office equipment.

ANALYSIS: The *assets* Cash and Office Equipment are increased, and the *owner's equity* account, Lloyd Thomas, Capital, is also increased by the same amount. The journal entries are

	Dr.	Cr.
Cash	$10,000	
Office Equipment	$ 5,000	
L. Thomas, Capital		$15,000

The related accounts would appear as follows:

Cash	I., Thomas, Capital
(1) 10,000	(1) 15,000

Office Equipment
(1)   5,000

Transaction 2. Acquired office supplies for cash, $800.

ANALYSIS: The asset Office Supplies goes up by $800 with a corresponding reduction in the asset Cash. This transaction represents the conversion of one asset to another. At the end of the month, supplies will be counted to determine the amount used during the month [see transaction (10)]. The entries are

	Dr.	Cr.
Cash	$800	
Office Supplies		$800

The related accounts would appear as follows:

Cash	Office Supplies
(1) 10,000	(2) 800
(2) 800	

Transaction 3. Purchased a word processor on account, $500.

ANALYSIS: An asset, Office Equipment, is being acquired on credit, thereby creating a *liability* for future payment called Accounts Payable. The entries are

	Dr.	Cr.
Office Equipment	$500	
Accounts Payable		$500

Office Equipment	Accounts Payable
(1) 5,000	(3) 500
(3) 500	

Transaction 4. Paid salary, $400.

ANALYSIS: The transaction reduces Cash and increases Wage Expense. When financial statements are prepared at the end of January, the month's wages will appear on the income statement as an expense. The entries are

	Dr.	Cr.
Wage Expense	$400	
Cash		$400

Wage Expense	Cash
(4) 400	(1) 10,000 (2) 800
	(4) 400

Transaction 5. Received an electric bill for $300 (not paid).

ANALYSIS: Liabilities are increased by $300 since the firm *owes* the utility money for electricity supplied. Since the utility service was used in January, this amount is a January expense and will be reflected in January's income statement. The entries are

	Dr.	Cr.
Utilities Expense	$300	
Accounts Payable		$300

Utilities Expense	Accounts Payable
(5) 300	(3) 500
	(5) 300

Transaction 6. Earned fees of $20,000, of which $12,000 was received in cash and $8,000 was owed by clients.

ANALYSIS: Cash goes up by $12,000 and the account Accounts Receivable (amounts owed to the business from customers) is created. Professional fees earned are *revenue* to the business and hence increases the owner's equity. Note that the $20,000 revenue is reflected in the month that service is given, even though clients may not pay $8,000 until a later period. The entries are

	Dr.	Cr.
Cash	$12,000	
Accounts Receivable	$8,000	
Professional Fee		$20,000

Cash				Professional Fee	
(1) 10,000	(2)	800		(6) 20,000	
(6) 12,000	(4)	400			

Accounts Receivable	
(6) 8,000	

Transaction 7. Paid $300 to Smith Corona (in partial payment of the amount owed to them).

ANALYSIS: The payment lowers the asset Cash and reduces the liability Accounts Payable. This payment is the partial settlement of a previously recorded obligation, not an expense. The entries are

	Dr.	Cr.
Accounts Payable	$300	
Cash		$300

Accounts Payable				Cash		
(7) 300	(3)	500		(1) 10,000	(2)	800
	(5)	300		(6) 12,000	(4)	400
					(7)	300

Transaction 8. Withdrew $100 for personal use.

ANALYSIS: Cash is reduced and so is equity. The personal withdrawal is, in effect, a disinvestment in the business and hence reduces equity. It is not an expense in running the business. The entries are

	Dr.	Cr.
L. Thomas, Capital	$100	
Cash		$100

L. Thomas, Capital	Cash
(8) 100 (1) 15,000	(1) 10,000 (2) 800
	(6) 12,000 (4) 400
	(7) 300
	(8) 100

Transaction 9. Received $1,000 from clients who owed him money.

ANALYSIS: This increases Cash and reduces Accounts Receivable since the client now owes the business less money. One asset is being substituted for another one. The entries are

	Dr.	Cr.
Cash	$1,000	
Accounts Receivable		$1,000

Cash	Accounts Receivable
(1) 10,000 (2) 800	(6) 8,000 (9) 1,000
(6) 12,000 (4) 400	
(9) 1,000 (7) 300	
(8) 100	

Transaction 10. Worth (determined by an inventory count) of office supplies on hand at month's end, $600.

ANALYSIS: Since the worth of office supplies originally acquired was $800 and $600 is left on hand, the business used $200 in supplies. This reduces the asset Office Supplies and correspondingly reduces equity. The supplies used up represent office supplies expense. The entries are

	Dr.	Cr.
Office Supplies Expense	$200	
Office Supplies		$200

Office Supplies Expense	Office Supplies
(10) 200	(1) 800 (10) 200

After the foregoing transactions have been entered properly, the account balances can be determined. The accounts are shown in Figures 3.1 and 3.2,

| Assets | = | Liabilities | + | Owner's Equity |

Cash

(1) $10,000	(2)	800
(6) 12,000	(4)	400
(9) 1,000	(7)	300
	(8)	100
Bal. 21,400		

Accounts Receivable

| (6) 8,000 | (9) | 1,000 |
| Bal. 7,000 | | |

Office Equipment

(1) 5,000	
(3) 500	
Bal. 5,500	

Office Supplies

| (1) 800 | (10) | 200 |
| Bal. 600 | | |

Accounts Payable

(7)	300	(3)	500
		(5)	300
		Bal.	500

L. Thomas, Capital

| (8) | 100 | (1) | 15,000 |
| | | Bal. | 14,900 |

Professional Fee

| | (6) | 20,000 |

Wage Expense

| (4) | 400 |

Utilities Expense

| (5) | 300 |

Office Supplies Expense

| (10) | 200 |

Figure 3.1 The accounts of Lloyd Thomas, a consulting engineer.

Lloyd Thomas
Balance Sheet
January 31, 19x1

Assets		*Liabilities*	
Cash	$ 21,400	Accounts Payable	$ 500
Accounts Receivable	7,000		
Office Equipment	5,500	*Owner's Equity*	
Office Supply	7,000	Capital:	
		Balance, Jan. 1	$ 0
		Add:	
		Contribution	15,000
		Net income, for Jan.	19,100
		Less: Withdrawals	100
		Balance, Jan. 31	34,000
		Total Liabilities	
Total assets	$ 34,500	and owner's equity	$ 34,500

Lloyd Thomas
Income Statement
For the Month Ended January 31, 19x1

Revenue:		
Professional Services		$20,000
Less: Expenses		
Wage Expense	$ 400	
Utilities Expense	300	
Office Supplies Expense	200	
Total Expenses		900
Net Income		$19,100

Figure 3.2 Lloyd Thomas' financial statements.

together with the financial statements prepared from the balances of these accounts.

Journals

For simplicity, the entries used in the previous section were made directly in the general ledger accounts. However, this process does not furnish the data required about a given transaction nor is the listing of transactions in chronological order

General Journal				Page No.
Date	Accounts	P.R.	Debit	Credit
Jan. 1	Cash		10,000	
	Office Equipment		5,000	
	L. Thomas, Capital			15,000
	Started his practice			
Jan. 8				

Figure 3.3 General journal entries.

possible on T-accounts. These deficiencies are overcome through the use of a *journal*. The journal is the book of original entry in which transactions are entered on a daily basis in chronological order. This process is called *journalizing*, as shown in Figure 3.3.

The data are then transferred from the journal to the ledger by debiting and crediting the particular accounts involved. This process is called *posting*. The P.R. (Posting Reference) column is used for the ledger account number after the posting from the journal to the ledger takes place. This provides a cross reference between journal and ledger. There exist different types of journals that may be grouped into the categories of (1) general journals and (2) specialized journals. The latter is used when there are many repetitive transactions (e.g., sales or payroll).

TYPES OF DEPRECIATION METHODS

Depreciation is an important issue in accounting policy and financial management. It requires an explicit discussion since its treatment has a considerable bearing on cash, reported earnings, tax liabilities incurred, and facilities investment decisions.

Depreciation is the decline in economic potential of fixed assets (with the exception of land) originating from wear and tear, deterioration, and obsolescence. Accounting for depreciation involves the process of spreading the cost of an asset over its useful life.

Among the commonly used depreciation methods are straight-line and accelerated methods. The two major accelerated methods are sum-of-the-years'-digits (SYD) and double-declining-balance (DDB). Each of these methods is explained below.

Straight-Line Method

This is the easiest and most popular method of calculating depreciation. It results in equal periodic depreciation charges. The method is most appropriate when an asset's usage is uniform from period to period, as is the case with furniture. The annual depreciation expense is calculated by using the following formula:

$$\text{Depreciation expense} = \frac{\text{Cost} - \text{Salvage value}}{\text{Number of years of useful life}}$$

Example 3.3. An auto is purchased for $20,000 and has an expected salvage value of $2,000. The auto's estimated life is 8 years. Its annual depreciation is calculated as follows:

$$\text{Depreciation expense} = \frac{\text{Cost} - \text{Salvage value}}{\text{Number of years of useful life}}$$

$$= \frac{\$20,000 - \$2,000}{8 \text{ years}} = \$2,250/\text{year}$$

An alternative means of computation is to multiply the *depreciable* cost ($18,000) by the annual depreciation rate, which is 12.5% in this example. The annual rate is calculated by dividing the number of years of useful life into one (1/8 = 12.5%). The result is the same: $18,000 × 12.5% = $2,250.

Sum-of-the-Years'-Digits (SYD) Method

In this method, the number of years of life expectancy is enumerated in reverse order in the numerator, and the denominator is the sum of the digits. For example, if the life expectancy of a machine is 8 years, write the numbers in reverse order: 8, 7, 6, 5, 4, 3, 2, 1. The sum of these digits is 36, or (8 + 7 + 6 + 5 + 4 + 3 + 2 + 1). Thus, the fraction for the first year is 8/36, while the fraction for the last year is 1/36. The sum of the eight fractions equals 36/36, or 1. Therefore, at the end of 8 years, the machine is completely written down to its salvage value.

The following formula may be used to quickly find the sum-of-the-years' digits (S):

$$S = \frac{(N)\,(N+1)}{2}$$

where N represents the number of years of expected life.

Example 3.4. In Example 3.3, the *depreciable* cost is $18,000 ($20,000 - $2,000). Using the SYD method, the computation for each year's depreciation expense is

$$S = \frac{(N)\,(N+1)}{2} - \frac{8(9)}{2} = \frac{72}{2} = 36$$

Year	Fraction × Depreciation amount ($)		= Depreciation expense
1	8/36	$18,000	$ 4,000
2	7/36	18,000	3,500
3	6/36	18,000	3,000
4	5/36	18,000	2,500
5	4/36	18,000	2,000
6	3/36	18,000	1,500
7	2/36	18,000	1,000
8	1/36	18,000	500
Total			$18,000

Double-Declining-Balance (DDB) Method

Under this method, depreciation expense is highest in the earlier years and lower in the later years. First, a depreciation rate is determined by doubling the straight-line rate. For example, if an asset has a life of 10 years, the straight-line rate is 1/10 or 10%, and the double-declining rate is 20%. Second, depreciation expense is computed by multiplying the rate by the book value of the asset at the beginning of each year. Since book value declines over time, the depreciation expense decreases each successive period.

This method *ignores* salvage value in the computation. However, the book value of the fixed asset at the end of its useful life cannot be below its salvage value.

Example 3.5. Review the data in Example 3.3. Since the straight-line rate is 12.5% (1/8), the double-declining-balance rate is 25% (2 × 12.5%). The depreciation expense is computed as follows:

Year	Book value at beginning of year	× Rate (%)	= Depreciation expense	Year-end book value
1	$20,000	25	$5,000	$15,000
2	15,000	25	3,750	11,250
3	11,250	25	2,813	8,437
4	8,437	25	2,109	6,328
5	6,328	25	1,582	4,746
6	4,746	25	1,187	3,559
7	3,559	25	890	2,669
8	2,669	25	667	2,002

Note: If the original estimated salvage value had been $2,100 instead of $2,000, the depreciation expense for the eighth year would have been $569 ($2,669 - $2,100) rather than $667, since the asset cannot be depreciated below its salvage value.

Which Method to Use

1. Of course, over the life of the fixed asset, the total depreciation charge will be the same no matter what depreciation method is used; only the timing of the tax savings will differ.
2. The accelerated methods such as SYD and DDB are advantageous for tax purposes since higher depreciation charges in the earlier years result in less income and thus less taxes. The tax savings may then be invested for a return.

CONCLUSION

Financial decisions are usually based on the basis of accounting information generated by the accounting system. This chapter introduced a formal classification system, known as double-entry accounting. To facilitate a basic understanding of this system, various accounting tools and concepts were covered with numerical examples. The depreciation policies adopted by an entity have significant implications for reported earnings, tax liabilities, cash positions, and capital investment decisions.

4

ANALYZING FINANCIAL STATEMENTS

This chapter covers how to analyze a company's financial statements which include the balance sheet and income statement. Financial statement analysis attempts to answer the following basic questions:

1. How well is the business doing?
2. What are its strengths?
3. What are its weaknesses?
4. How does it fare in the industry?
5. Is the business improving or deteriorating?

A complete set of financial statements, as explained in the previous chapter, will include the balance sheet, income statement, and statement of cash flows. The first two are vital in financial statement analysis. We will discuss the various financial statement analysis tools that you will use in evaluating the firm's present and future financial condition. These tools include horizontal, vertical, and ratio analysis, which give relative measures of the performance and financial condition of the company.

WHAT AND WHY OF FINANCIAL STATEMENT ANALYSIS

The analysis of financial statements means different things to different people. It is of interest to creditors, present and prospective investors, and the firm's own management.

A *creditor* is primarily interested in the firm's debt-paying ability. A short-term creditor, such as a vendor or supplier, is ultimately concerned with the firm's ability to pay its bills and therefore wants to be assured that the firm is liquid. A long-term creditor such as a bank or bondholder on the other hand, is interested in the firm's ability to repay interest and principal on borrowed funds.

An *investor* is interested in the present and future level of return (earnings) and risk (liquidity, debt, and activity). You, as an investor, must evaluate a firm's stock based on an examination of its financial statements. This evaluation considers overall financial health, economic and political conditions, industry factors, and future outlook of the company. The analysis attempts to ascertain whether the stock is overpriced, underpriced, or priced in proportion to its market value. A stock is valuable to you only if you can predict the future financial performance of the business. Financial statement analysis gives you much of the data you will need to forecast earnings and dividends.

Management must relate the analysis to all of the questions raised by creditors and investors, since these interested parties must be satisfied for the firm to obtain capital as needed.

HORIZONTAL AND VERTICAL ANALYSIS

Comparison of two or more years' financial data is known as *horizontal analysis*. Horizontal analysis concentrates on the trend in the accounts in dollar *and* percentage terms over the years. It is typically presented in comparative financial statements (see Biogen, Inc. financial data in Figures 4.1 and 4.2). In annual reports, comparative financial data are usually shown for five years.

Through horizontal analysis you can pinpoint areas of wide divergence requiring investigation. For example, in the income statement shown in Figure 4.2, the significant rise in sales returns taken with the reduction in sales for 19x1-19x2 should cause concern. You might compare these results with those of competitors.

It is essential to present both the dollar amount of change and the percentage of change since the use of one without the other may result in erroneous conclusions. The interest expense from 19x0-19x1 went up by 100.0%, but this represented only $1,000 and may not need further investigation. In a similar vein, a large number change might cause a small percentage change and not be of any great importance.

Key changes and trends can also be highlighted by the use of *common-size statements*. A common size statement is one that shows the separate items appearing on it in percentage term. Preparation of common-size statements is known as *vertical analysis*. In vertical analysis, a material financial statement item is used as a base value, and all other accounts on the financial statement are compared to it. In the balance sheet, for example, total assets equal 100%. Each asset is stated as a percentage of total assets. Similarly, total liabilities and stockholders' equity is assigned 100% with a given liability or equity account stated as a percentage of the total liabilities and stockholders' equity.

Biogen Inc.
Comparative Balance Sheet (In Thousands of Dollars)
December 31, 19X2, 19X1, 19X0

	19X2	19X1	19X0	Incr. or Decr. 19X2-19X1	Decr. 19X1-19X0	% Incr. 19x2-19X1	or Decr. 19X1-19X0
ASSETS							
Current Assets:							
Cash	$28	$36	$36	-8	0	-22.2%	0.0%
Marketable Securities	$22	$15	$7	7	8	46.7%	114.3%
Accounts Receivable	$21	$16	$10	5	6	31.3%	60.0%
Inventory	$53	$46	$49	7	-3	15.2%	-6.1%
Total Current Assets	$124	$113	$102	11	11	9.7%	10.8%
Plant And Equip.	$103	$91	$83	12	8	13.2%	9.6%
Total Assets	$227	$204	$185	23	19	11.3%	10.3%
LIABILITIES							
Current Liabilities	$56	$50	$51	6	-1	12.0%	-2.0%
Long-term debt	$83	$74	$69	9	5	12.2%	7.2%
Total Liabilities	$139	$124	$120	15	4	12.1%	3.3%
STOCKHOLDERS' EQUITY							
Common Stock, $10 par, 4,600 shares	$46	$46	$46	0	0	0.0%	0.0%
Retained Earnings	$42	$34	$19	8	15	23.5%	78.9%
Total Stockholders' Equity	$88	$80	$65	8	15	10.0%	23.1%
Total Liab.and Stockholders' Equity	$227	$204	$185	23	19	11.3%	10.3%

Figure 4.1 Comparative balance sheet.

Placing all assets in common-size form clearly shows the relative importance of the current assets as compared to the noncurrent assets. It also shows that significant changes have taken place in the composition of the current assets over the last year. Notice, for example, that receivables have increased in relative importance and that cash has declined in relative importance. The deterioration in the cash position may be a result of inability to collect from customers.

For the income statement, 100% is assigned to net sales with all other revenue and expense accounts related to it. It is possible to see at a glance how each dollar of sales is distributed between the various costs, expenses, and profits. For example, notice from Figure 4.3 that 64.8 cents of every dollar of sales were needed to cover cost of goods sold in 19x2, as compared to only 57.3 cents in the

Biogen Inc.
Comparative Income Statement (In Thousands of Dollars)
For the Years Ended December 31, 19X2, 19X1, 19X0

	19X2	19X1	19X0	Incr. or Decr. 19X2-19X1	Decr. 19X1-19X0	% Incr. 19X2-19X1	or Decr. 19X1-19X0
Sales	$98.3	$120.0	$56.6	($21.7)	$63.4	-18.1%	112.0%
Sales Return & Allowances	$18.0	$10.0	$4.0	$8.0	$6.0	80.0%	150.0%
Net Sales	$80.3	$110.0	$52.6	($29.7)	$57.4	-27.0%	109.1%
Cost of Goods Sold	$52.0	$63.0	$28.0	($11.0)	$35.0	-17.5%	125.0%
Gross Profit	$28.3	$47.0	$24.6	($18.7)	$22.4	-39.8%	91.1%
Operating Expenses							
Selling Expenses	$12.0	$13.0	$11.0	($1.0)	$2.0	-7.7%	18.2%
Administrative Expenses	$5.0	$8.0	$3.0	($3.0)	$5.0	-37.5%	166.7%
Total Operating Expenses	$17.0	$21.0	$14.0	($4.0)	$7.0	-19.0%	50.0%
Income from Operations	$11.3	$26.0	$10.6	($14.7)	$15.4	-56.5%	145.3%
Nonoperating Income	$4.0	$1.0	$2.0	$3.0	($1.0)	300.0%	-50.0%
Income before Interest & Taxes	$15.3	$27.0	$12.6	($11.7)	$14.4	-43.3%	114.3%
Interest Expense	$2.0	$2.0	$1.0	$0.0	$1.0	0.0%	100.0%
Income before Taxes	$13.3	$25.0	$11.6	($11.7)	$13.4	-46.8%	115.5%
Income Taxes (40%)	$5.3	$10.0	$4.6	($4.7)	$5.4	-46.8%	115.5%
Net Income	$8.0	$15.0	$7.0	($7.0)	$8.0	-46.8%	115.5%

Figure 4.2 Comparative income statement.

prior year; also notice that only 9.9 cents out of every dollar of sales remained for profits in 19x2—down from 13.6 cents in the prior year.

You should also compare the vertical percentages of the business to those of the competition and to the industry norms. Then you can determine how the company fares in the industry.

Figure 4.3 shows a common size income statement based on the data provided in Figure 4.2.

WORKING WITH FINANCIAL RATIOS

Horizontal and vertical analysis compares one figure to another within the same category. It is also vital to compare two figures applicable to different categories. This is accomplished by ratio analysis. In this section, you will learn how to

Biogen Inc.
Income Statement and Common Size Analysis (In Thousands of Dollars)
For the Years Ended December 31, 19X2 & 19X1

	19X2 Amount	%	19X1 Amount	%
Sales	$98.3	122.4%	$120.0	109.1%
Sales Return & Allowances	$18.0	22.4%	$10.0	9.1%
Net Sales	$80.3	100.0%	$110.0	100.0%
Cost of Goods Sold	$52.0	64.8%	$63.0	57.3%
Gross Profit	$28.3	35.2%	$47.0	42.7%
Operating Expenses				
Selling Expenses	$12.0	14.9%	$13.0	11.8%
Administrative Expenses	$5.0	6.2%	$8.0	7.3%
Total Operating Expenses	$17.0	21.2%	$21.0	19.1%
		0.0%		0.0%
Income from Operations	$11.3	14.1%	$26.0	23.6%
Nonoperating Income	$4.0	5.0%	$1.0	0.9%
Income before Interest & Taxes	$15.3	19.1%	$27.0	24.5%
Interest Expense	$2.0	2.5%	$2.0	1.8%
Income before Taxes	$13.3	16.6%	$25.0	22.7%
Income Taxes (40%)	$5.3	6.6%	$10.0	9.1%
Net Income	$8.0	9.9%	$15.0	13.6%

Figure 4.3 Income statement with analysis.

calculate the various financial ratios and how to interpret them. The results of the ratio analysis will allow you:

1. To appraise the position of a business
2. To identify trouble spots that need attention
3. To provide the basis for making projections and forecasts about the course of future operations

Think of ratios as measures of the relative health or sickness of a business. Just as a doctor takes readings of a patient's temperature, blood pressure, heart rate, etc., you will take readings of a business's liquidity, profitability, leverage, efficiency in using assets, and market value. Where the doctor compares the readings to generally accepted guidelines such as a temperature of 98.6 degrees as normal, you make some comparisons to the norms.

To obtain useful conclusions from the ratios, you must make two comparisons:

Industry comparison. This will allow you to answer the question "how does a business fare in the industry?" You must compare the company's ratios to those of competing companies in the industry or with industry standards (averages). You can obtain industry norms from financial services such as *Value Line, Dun and Bradstreet*, and *Standard and Poor's*.

Trend analysis. To see how the business is doing over time, you will compare a given ratio for one company over several years to see the direction of financial health or operational performance.

Financial ratios can be grouped into the following types: liquidity, asset utilization (activity), solvency (leverage and debt service), profitability, and market value.

Liquidity Ratios

Liquidity is the firm's ability to satisfy maturing short-term debt. Liquidity is crucial to carrying out the business, especially during periods of adversity. It relates to the short term, typically a period of one year or less. Poor liquidity might lead to higher cost of financing and inability to pay bills and dividends. The three basic measures of liquidity are (1) net working capital, (2) the current ratio, and (3) the quick (acid-test) ratio.

Throughout our discussion, keep referring to Figures 4.1 and 4.2 to make sure you understand where the numbers come from.

Net working capital equals current assets minus current liabilities. Net working capital for 19x2 is:

Net working capital = Current assets − Current liabilities

$$= \$124 \qquad - \$56$$

$$= \$68$$

In 19x1, net working capital was $63. The rise over the year is favorable.

The current ratio equals current assets divided by current liabilities. The ratio reflects the company's ability to satisfy current debt from current assets.

$$\text{Current ratio} = \frac{\text{Current assets}}{\text{Current liabilities}}$$

For 19x2, the current ratio is:

$$\frac{\$124}{\$56} = 2.21$$

In 19x1, the current ratio was 2.26. The ratio's decline over the year points to a slight reduction in liquidity.

A more stringent liquidity test can be found in the *quick (acid-test) ratio.* Inventory and prepaid expenses are excluded from the total of current assets, leaving only the more liquid (or quick) assets to be divided by current liabilities.

$$\text{Quick ratio} = \frac{\text{Cash + Marketable securities + Accounts receivable}}{\text{Current liabilities}}$$

The quick ratio for 19x2 is:

$$\frac{\$28 + \$21 + \$22}{\$56} = 1.27$$

In 19x1, the ratio was 1.34. A small reduction in the ratio over the period points to less liquidity.

The overall liquidity trend shows a slight deterioration as reflected in the lower current and quick ratios, although it is better than the industry norms (see Figure 4.4 for industry averages). But a mitigating factor is the increase in net working capital.

Summary of Financial Ratios
Trend and Industry Comparisons
Biogen, Inc. 19X2 and 19X1

Ratios	Definitions	19X1	19X2	(a) 19X2 Industry Average	19X2 Ind. Comp.	Evaluation (b) Trend 19X1-19X2	Overall
LIQUIDITY							
Net working capital	Current assets - current liabilities	63	68	56	good	good	good
Current Ratio	Current assets/current liabilities	2.26	2.21	2.05	OK	OK	OK
Quick (Acid-test) ratio	(Cash + marketable securities + accounts receivable)/current liabilities	1.34	1.27	1.11	OK	OK	OK
ASSET UTILIZATION							
Accounts receivable turnover	Net credit sales/average accounts receivable	8.46	4.34	5.5	OK	poor	poor
Average collection period	365 days/accounts receivable turnover	43.1 days	84.1 days	66.4 days	OK	poor	poor
Inventory turnover	Cost of goods sold/average inventory	1.33	1.05	1.2	OK	poor	poor

Average age of inventory	365 days/inventory turnover	274.4 days	347.6 days	N/A	poor	poor
Operating cycle	Average collection period + average age of inventory	317.5 days	431.7 days	N/A	poor	poor
Total asset turnover	Net sales/average total assets	0.57	0.37	0.44 OK	poor	poor
SOLVENCY						
Debt ratio	Total liabilities/total assets	0.61	0.61	N/A	OK	OK
Debt-equity ratio	Total liabilities/stockholders' equity	1.55	1.58	1.3 poor	poor	poor
Times interest earned	Income before interest and taxes/interest expense	13.5 times	7.65 times	13 times OK	poor	poor
PROFITABILITY						
Gross profit margin	Gross profit/net sales	0.43	0.35	0.43 poor	poor	poor
Profit margin	Net income/net sales	0.14	0.1	0.15 poor	poor	poor
Return on total assets	Net income/average total assets	0.077	0.037	0.1 poor	poor	poor

Figure 4.4 Summary of Financial Ratios, Trends, Comparisons.

Ratios	Definitions	19X1	19X2	(a) 19X2 Industry Average	19X2 Ind. Comp.	Evaluation (b) Trend 19X1-19X2	Overall
Return on equity (ROE)	Earnings available to common stockholders/ average stockholders' equity	0.207	0.095	0.27	poor	poor	poor
MARKET VALUE							
Earnings per share (EPS)	(Net income -preferred dividend)/ common shares outstanding	3.26	1.74	4.51	poor	poor	poor
Price/earnings (P/E) ratio	Market price per share/EPS	7.98	6.9	7.12	OK	poor	poor
Book value per share	(Total stockholders' equity - preferred stock)/ common shares outstanding	17.39	19.13	N/A	N/A	good	good
Price/book value ratio	Market price per share/book value per share	1.5	0.63	N/A	N/A	poor	poor
Dividend yield	Dividends per share/market price per share						
Dividend payout	Dividends per share/EPS						

(a) Obtained from sources not included in this chapter
(b) Represent subjective evaluation

Figure 4.4 Continued.

Asset Utilization Ratios

Asset utilization (activity, turnover) ratios reflect the way in which a company uses its assets to obtain revenue and profit. One example is how well receivables are turning into cash. The higher the ratio, the more efficiently the business manages its assets.

Accounts receivable ratios comprise the accounts receivable turnover and the average collection period.

The *accounts receivable turnover* provides the number of times accounts receivable are collected in the year. It is derived by dividing net credit sales by average accounts receivable.

You can calculate average accounts receivable by the average accounts receivable balance during a period.

$$\text{Accounts receivable turnover} = \frac{\text{Net credit sales}}{\text{Average accounts receivable}}$$

For 19x2, the average accounts receivable is:

$$\frac{\$21 + \$16}{2} = \$18.5$$

The accounts receivable turnover for 19x2 is:

$$\frac{\$80.3}{\$18.5} = 4.34$$

In 19x1, the turnover was 8.46. There is a sharp reduction in the turnover rate pointing to a collection problem.

The *average collection period* is the length of time it takes to collect receivables. It represents the number of days receivables are held.

$$\text{Average collection period} = \frac{365 \text{ days}}{\text{Accounts receivable turnover}}$$

In 19x2, the collection period is:

$$\frac{365}{4.34} = 84.1 \text{ days}$$

It takes this firm about 84 days to convert receivables to cash. In 19x1, the collection period was 43.1 days. The significant lengthening of the collection period may be a cause for some concern. The long collection period may be a result of the presence of many doubtful accounts, or it may be a result of poor credit management. *Inventory ratios* are useful especially when a buildup in inventory exists. Inventory ties up cash. Holding large amounts of inventory can result in lost opportunities for profit as well as increased storage costs. Before you extend credit or lend money, you should examine the firm's *inventory turnover* and *average age of inventory*.

$$\text{Inventory turnover} = \frac{\text{Cost of goods sold}}{\text{Average inventory}}$$

The inventory turnover for 19x2 is:

$$\frac{\$52}{\$49.5} = 1.05$$

For 19x1, the turnover was 1.33.

$$\text{Average age of inventory} = \frac{365}{\text{Inventory turnover}}$$

In 19x2, the average age is:

$$\frac{365}{1.05} = 347.6 \text{ days}$$

In the previous year, the average age was 274.4 days.

The reduction in the turnover and increase in inventory age points to a longer holding of inventory. You should ask why the inventory is not selling as quickly.

The *operating cycle* is the number of days it takes to convert inventory and receivables to cash.

Operating cycle = Average collection period + Average age of inventory

In 19x2, the operating cycle is:

84.1 days + 347.6 days = 431.7 days

In the previous year, the operating cycle was 317.5 days. An unfavorable direction is indicated because additional funds are tied up in noncash assets. Cash is being collected more slowly.

By calculating the *total asset turnover*, you can find out whether the company is efficiently employing its total assets to obtain sales revenue. A low ratio may indicate too high an investment in assets in comparison to the sales revenue generated.

$$\text{Total asset turnover} = \frac{\text{Net sales}}{\text{Average total assets}}$$

In 19x2, the ratio is:

$$\frac{\$80.3}{(\$204 + \$227)/2} = \frac{\$80.3}{\$215.5} = .37$$

In 19x1, the ratio was .57 ($110/$194.5). There has been a sharp reduction in asset utilization.

Biogen, Inc. has suffered a sharp deterioration in activity ratios, pointing to a need for improved credit and inventory management, although the 19x2 ratios are not far out of line with the industry averages (Figure 4.4). It appears that problems are inefficient collection and obsolescence of inventory.

Solvency (Leverage and Debt Service) Ratios

Solvency is the company's ability to satisfy long-term debt as it becomes due. You should be concerned about the long-term financial and operating structure of any firm in which you might have a vested interest. Another important consideration is the size of debt in the firm's capital structure, which is referred to as *financial leverage*. (Capital structure is the mix of the *long term* sources of funds used by the firm.)

Solvency also depends on earning power; in the long run a company will not satisfy its debts unless it earns profit. A leveraged capital structure subjects the company to fixed interest charges, which contributes to earnings instability. Excessive debt may also make it difficult for the firm to borrow funds at reasonable rates during tight money markets.

The *debt ratio* reveals the amount of money a company owes to its creditors. Excessive debt means greater risk to the investor (note that equity holders come after creditors in bankruptcy.) The debt ratio is:

$$\frac{\text{Total liabilities}}{\text{Total assets}}$$

In 19x2, the ratio is:

$$\frac{\$139}{\$227} = .61$$

The *debt-equity ratio* will show you if the firm has a great amount of debt in its capital structure. Large debts mean that the borrower has to pay significant periodic interest and principal. Also, a heavily indebted firm takes a greater risk of running out of cash in difficult times. The interpretation of this ratio depends on several variables, including the ratios of other firms in the industry, the degree of access to additional debt financing, and stability of operations.

The ratio equals:

$$\frac{\text{Total liabilities}}{\text{Stockholders' equity}}$$

In 19x2, the ratio is:

$$\frac{\$139}{\$88} = 1.58$$

In the previous year, the ratio was 1.55. The trend is relatively static.

Times interest earned (interest coverage ratio) tells you how many times the firm's before-tax earnings would cover interest. It is a safety margin indicator in that it reflects how much of a reduction in earnings a company can tolerate.

$$\frac{\text{Income before interest and taxes}}{\text{Interest expense}}$$

For 19x2, the ratio is:

$$\frac{\$15.3}{\$2.0} = 7.56$$

In 19x1, interest was covered 13.5 times. The reduction in coverage during the period is a bad sign. It means that less earnings are available to satisfy interest charges.

You must also note liabilities that have not yet been reported in the balance sheet by closely examining footnote disclosure. For example, you should find out about lawsuits, noncapitalized leases, and future guarantees.

As shown in Figure 4.4, the company's overall solvency is poor, relative to the industry averages although it has remained fairly constant. There has been no significant change in its ability to satisfy long-term debt. Note that significantly less profit is available to cover interest payments.

Profitability Ratios

A company's ability to earn a good profit and return on investment is an indicator of its financial well-being and the efficiency with which it is managed. Poor earnings have detrimental effects on the market price of stock and dividends. Total dollar net income has little meaning unless it is compared to the input in getting that profit.

The *gross profit margin* shows the percentage of each dollar remaining once the company has paid for goods acquired. A high margin reflects good earning potential.

$$\text{Gross profit margin} = \frac{\text{Gross profit}}{\text{Net sales}}$$

In 19x2, the ratio is:

$$\frac{\$28.3}{\$80.3} = .35$$

The ratio was .43 in 19x1. The reduction shows that the company now receives less profit on each dollar of sales. Perhaps higher relative cost of merchandise sold is at fault.

Profit margin shows the earnings generated from revenue and is a key indicator of operating performance. It gives you an idea of the firm's pricing, cost structure, and production efficiency.

$$\text{Profit margin} = \frac{\text{Net income}}{\text{Net sales}}$$

The ratio in 19x2 is:

$$\frac{\$8}{\$80.3} = .10$$

For the previous year, profit margin was .14. The decline in the ratio shows a downward trend in earning power. (Note that these percentages are available in the common size income statement as given in Figure 4.3).

Table 4.1 shows various profit margin ratios.

Table 4.1 Profit Margin

As of 2nd Quarter, 1993	
Burlington Resources	42.8%
Amgen	29.2
H&R Block	22.5
Intel	26.7
Albertson's	2.7
Campbell Soup	6.6

Source: Business Week, *Industry Scorecard*,
McGraw-Hill, August 16, 1993.

Return on investment is a prime indicator because it allows you to evaluate the profit you will earn if you invest in the business. Two key ratios are the *return on total assets* and the *return on equity*.

The return on total assets shows whether management is efficient in using available resources to get profit.

$$\text{Return on total assets} = \frac{\text{Net income}}{\text{Average total assets}}$$

In 19x2, the return is:

$$\frac{\$8}{(\$227 + \$204)/2} = .037$$

In 19x1, the return was .077. There has been a deterioration in the productivity of assets in generating earnings.

The *return on equity* (ROE) reflects the rate of return earned on the stockholders' investment.

$$\text{Return on common equity} = \frac{\text{Net income}}{\text{Average stockholders' equity}}$$

The return in 19x2 is:

$$\frac{\$8}{(\$88 + \$80)/2} = 0.095$$

In 19x1, the return was .207. There has been a significant drop in return to the owners.

Table 4.2 Companies with High Return on
Equity (ROE) Rates (in Excess of 30%)

12 Months Ending 6/30/93	
Carter Hawley	237.5%
Avon Products	130.3
UST	73.6
Neiman Marcus	64.4
Coca-Cola	52.1
General Mills	41.5
Conseco	40.1
Microsoft	31.7

Source: Business Week, *Industry Scorecard*,
McGraw-Hill, August 16, 1993.

The overall profitability of the company has decreased considerably, causing
a decline in both the return on assets and return on equity. Perhaps lower earnings were due in part to higher costs of short-term financing arising from the
decline in liquidity and activity ratios. Moreover, as turnover rates in assets go
down, profit will similarly decline because of a lack of sales and higher costs of
carrying higher current asset balances. As indicated in Figure 4.4, industry comparisons reveal that the company is faring very poorly in the industry.

Table 4.2 shows companies with high return on equity (in excess of 30%).

Market Value Ratios

Market value ratios relate the company's stock price to its earnings (or book
value) per share. Also included are dividend-related ratios.

Earnings per share (EPS) is the ratio most widely watched by investors.
EPS shows the net income per common share owned. You must reduce net
income by the preferred dividends to obtain the net income available to common
stockholders. Where preferred stock is not in the capital structure, you determine
EPS by dividing net income by common shares outstanding. EPS is a gauge of
corporate operating performance and of expected future dividends.

$$\text{EPS} = \frac{\text{Net income} - \text{Preferred dividend}}{\text{Common shares outstanding}}$$

EPS in 19x2 is:

$$\frac{\$8,000}{4,600 \text{ shares}} = \$1.74$$

For 19x1, EPS was $3.26. The sharp reduction over the year should cause alarm among investors. As you can see in Figure 4, the industry average EPS in 19x2 is much higher than that of Biogen, Inc. ($4.51 per share vs. $1.74 per share).

The *price/earnings (P/E) ratio*, also called *earnings multiple*, reflects the company's relationship to its stockholders. The P/E ratio represents the amount investors are willing to pay for each dollar of the firm's earnings. A high multiple (cost per dollar of earnings) is favored since it shows that investors view the firm positively. On the other hand, investors looking for value would prefer a relatively lower multiple (cost per dollar of earnings) as compared with companies of similar risk and return.

$$\text{P/E ratio} = \frac{\text{Market price per share}}{\text{Earnings per share}}$$

Assume a market price per share of $12 on December 31, 19x2, and $26 on December 31, 19x1. The P/E ratios are:

$$19\text{x}2: \frac{\$12}{\$1.74} = 6.9$$

$$19\text{x}1: \frac{\$26}{\$3.26} = 7.98$$

From the lower P/E multiple, you can infer that the stock market now has a lower opinion of the business. However, some investors argue that a low P/E ratio can mean that the stock is undervalued. Nevertheless, the decline over the year in stock price was 54% ($14/$26), which should cause deep investor concern.

Table 4.3 shows price-earnings ratios of certain companies.

Table 4.3 P/E Ratios

Company	Industry	As of 7/23/93
Apple	Computer	17
Goodrich	Tire	55
Weyerhauser	Forest products	15
GTE	Communications	18
Dow Chemical	Chemical	36

Source: Business Week, *Industry Scorecard*,
McGraw-Hill, August 16, 1993.

Book value per share equals the net assets available to common stockholders divided by shares outstanding. By comparing it to market price per share you can get another view of how investors feel about the business.

The book value per share in 19x2 is:

$$\frac{\text{Total stockholders' equity} - \text{Preferred stock}}{\text{Common shares outstanding}}$$

$$\frac{\$88,000 - 0}{4,600} = \$19.13$$

In 19x1, book value per share was $17.39.

The increased book value per share is a favorable sign, because it indicates that each share now has a higher book value. However, in 19x2, market price is much less than book value, which means that the stock market does not value the security highly. In 19x1, market price did exceed book value, but there is now some doubt in the minds of stockholders concerning the company. However, some analysts may argue that the stock is underpriced.

The *price/book value ratio* shows the market value of the company in comparison to its historical accounting value. A company with old assets may have a high ratio whereas one with new assets may have a low ratio. Hence, you should note the changes in the ratio in an effort to appraise the corporate assets.

The ratio equals:

$$\frac{\text{Market price per share}}{\text{Book value per share}}$$

In 19x2, the ratio is:

$$\frac{\$12}{\$19.13} = .63$$

In 19x1, the ratio was 1.5. The significant drop in the ratio may indicate a lower opinion of the company in the eyes of investors. Market price of the stock may have dropped because of a deterioration in liquidity, activity, and profitability ratios. The major indicators of a company's performance are intertwined (i.e., one affects the other) so that problems in one area may spill over into another. This appears to have happened to the company in our example.

Dividend ratios help you determine the current income from an investment. Two relevant ratios are:

Table 4.4 Dividend Yield and Payout Ratios

| | 1992 | |
	Yield	Payout
The Limited	1.10%	22%
Texaco	5.08	91
Exxon	4.55	75
Mattel	0.90	14
AT&T	2.34	46

Source: Business Week, *Industry Scorecard*,
McGraw-Hill, August 16, 1993.

$$\text{Dividend yield} = \frac{\text{Dividends per share}}{\text{Market price per share}}$$

$$\text{Dividend payout} = \frac{\text{Dividend per share}}{\text{Earnings per share}}$$

Table 4.4 shows the dividend yield and payout ratios of some companies.

There is no such thing as a "right" payout ratio. Stockholders look unfavorably upon reduced dividends because it is a sign of possible deteriorating financial health. However, companies with ample opportunities for growth at high rates of return on assets tend to have low payout ratios.

An Overall Evaluation—Summary of Financial Ratios

As indicated in this chapter, a single ratio or a single group of ratios is not adequate for assessing all aspects of the firm's financial condition. Figure 4.4 summarizes the 19x1 and 19x2 ratios calculated in the previous sections, along with the industry average ratios for 19x2. The figure also shows the formula used to calculate each ratio. The last three columns of the figure contain subjective assessments of Biogen's financial condition, based on trend analysis and 19x2 comparisons to the industry norms. (5-year ratios are generally needed for trend analysis to be more meaningful, however.)

By appraising the trend in the company's ratios from 19x1 to 19x2, we see from the drop in the current and quick ratios that there has been a slight detraction in short-term liquidity, although they have been above the industry averages. However, working capital has improved. A material deterioration in the activity ratios has occurred, indicating that improved credit and inventory policies are required. They are not terribly alarming, however, because these ratios are not way out of line with industry averages. Also, total utilization of assets, as indicated by the total asset turnover, shows a deteriorating trend.

Leverage (amount of debt) has been constant. However, there is less profit available to satisfy interest charges. Biogen's profitability has deteriorated over the year. In 19x2, it is consistently below the industry average in every measure of profitability. As a result, the return on the owner's investment and the return on total assets have gone down. The earnings decrease may be partly due to the firm's high cost of short-term financing and partly due to operating inefficiency. The higher costs may be due to receivable and inventory difficulties that forced a decline in the liquidity and activity ratios. Furthermore, as receivables and inventory turn over less, profit will fall off from a lack of sales and the costs of carrying more in current asset balances.

The firm's market value, as measured by the price/earnings (P/E) ratio, is respectable as compared with the industry. But it shows a declining trend.

In summary, it appears that the company is doing satisfactorily in the industry in many categories. The 19x1-19x2 period, however, seems to indicate that the company is heading for financial trouble in terms of earnings, activity, and short-term liquidity. The business needs to concentrate on increasing operating efficiency and asset utilization.

IS RATIO ANALYSIS A PANACEA?

While ratio analysis is an effective tool for assessing a business's financial condition, you must also recognize the following limitations:

1. Accounting policies vary among companies and can inhibit useful comparisons. For example, the use of different depreciation methods (straight-line vs. double declining balance) will affect profitability and return ratios.
2. Management may "fool around" with ("window-dress") the figures. For example, it can reduce needed research expense just to bolster net income. This practice, however, will almost always hurt the company in the long run.
3. A ratio is static and does not reveal future flows. For example, it will not answer questions such as "How much cash do you have in your pocket now?" or "Is that sufficient, considering your expenses and income over the next month?"
4. A ratio does not indicate the quality of its components. For example, a high quick ratio may contain receivables that may not be collected.
5. Reported liabilities may be undervalued. An example is a lawsuit on which the company is contingently liable.

6. The company may have multiple lines of business, making it difficult to identify the industry group the company is a part.
7. Industry averages cited by financial advisory services are only approximations. Hence, you may have to compare a company's ratios to those of competing companies in the industry.

CONCLUSION

The analysis of financial statements means different things to different people. It is of interest to creditors, present and prospective investors, and the firm's own management. This chapter has presented the various financial statement analysis tools useful in evaluating the firm's present and future financial condition. These techniques include horizontal, vertical, and ratio analysis, which provide relative measures of the performance and financial health of the company. Two methods were demonstrated for analyzing financial ratios. The first involved trend analysis for the company over time; the second involved making comparisons with industry norms.

While ratio analysis is an effective tool for assessing a company's financial condition, the limitation of ratios must be recognized.

5

COST CONCEPTS, COST BEHAVIOR, AND COST ACCOUNTING

In financial accounting, the term "cost" is defined as a measurement, in monetary terms, of the amount of resources used for some purpose. In managerial accounting, the term "cost" is used in many different ways. That is, there are different types of costs used for different purposes. Some costs are useful and required for inventory valuation and income determination. Some costs are useful for planning, budgeting, and cost control. Still others are useful for making short-term and long-term decisions.

Engineers and engineering managers must have a general understanding of cost accumulation, analysis, and reporting because they must be familiar with how costs associated with their responsibility units are determined. A basic knowledge of different types of costs is needed for cost control and selling price determination.

The objectives of the chapter are:

To identify and give examples of each of the basic cost elements involved in the manufacture of the product.

To explain and illustrate various cost concepts.

To explain the difference between the financial statements of a manufacturer and those of a merchandising firm.

To distinguish between variable, fixed, and mixed costs and explain the difference in their behavior.

To discuss how mixed costs are separated and how a cost-volume formula is used for flexible budgeting.

To understand two primary cost accumulation systems—job order costing and process costing.

COST CLASSIFICATION

Costs can be classified into various categories, according to:

1. Their management function
   a. Manufacturing costs
   b. Nonmanufacturing (operating) costs
2. Their ease of traceability
   a. Direct costs
   b. Indirect costs
3. Their timing of charges against sales revenue
   a. Product costs
   b. Period costs
4. Their behavior in accordance with changes in activity
   a. Variable costs
   b. Fixed costs
   c. Mixed (semivariable) costs
5. Their degree of averaging
   a. Total costs b. Unit (average) costs
6. Their relevance to planning, control and decision making
   a. Controllable and noncontrollable costs
   b. Standard costs
   c. Incremental costs
   d. Sunk costs
   e. Out-of-pocket costs
   f. Relevant costs
   g. Opportunity costs

Costs by Management Function

In a manufacturing firm, costs are divided into two major categories, by the functional activities they are associated with: (1) manufacturing costs, and (2) nonmanufacturing costs, also called operating expenses.

Manufacturing Costs

Manufacturing costs are those costs associated with the manufacturing activities of the company. Manufacturing costs are subdivided into three categories: direct materials, direct labor, and factory overhead. Direct materials are all materials that become an integral part of the finished product. Examples are the steel used to make an automobile and the wood used to make furniture. Glues, nails, and other minor items are called indirect materials (or supplies) and are classified as part of factory overhead, which is explained below.

Direct labor is the labor directly involved in making the product. Examples of direct labor costs are the wages of assembly workers on an assembly line and

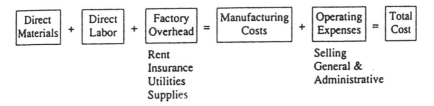

Figure 5.1 Costs by Management Function.

the wages of machine tool operators in a machine shop. Indirect labor, such as wages of supervisory personnel and janitors, is classified as part of factory overhead. Factory overhead can be defined as including all costs of manufacturing except direct materials and direct labor. Some of the many examples include depreciation, rent, taxes, insurance, fringe benefits, payroll taxes, and cost of idle time. Factory overhead is also called *manufacturing overhead, indirect manufacturing expenses, and factory burden.* Many costs overlap within these categories. For example, direct materials and direct labor when combined are called *prime costs.* Direct labor and factory overhead when combined are termed *conversion costs* (or processing costs).

Nonmanufacturing Costs

Nonmanufacturing costs (or operating expenses) are subdivided into selling expenses and general and administrative expenses. Selling expenses are those associated with obtaining sales and the delivery of the product. Examples are advertising and sales commissions. General and administrative expenses (G & A) include those incurred to perform general and administrative activities. Examples are executives' salaries and legal expenses. Other examples of costs by management function and their relationships are found in Figure 5.1.

Direct Costs and Indirect Costs

Costs may be viewed as either direct or indirect in terms of the extent that they are traceable to particular objects of costing such as products, jobs, departments, and sales territories. Direct costs can be directly traceable to the costing object. For example, if the object of costing under consideration is a product line, then the materials and labor involved in the manufacture of the line would both be direct costs. Factory overhead items are all indirect costs since they are not directly identifiable with any particular product line. Costs shared by different departments, products, or jobs, called *common costs* or *joint costs*, are also indirect costs. National advertising that benefits more than one product and sales territory is an example of an indirect cost.

Product Costs and Period Costs

By the timing of charges against revenue or by whether they are inventoriable, costs are classified into (a) product costs, and (b) period costs.

Product costs are inventoriable costs, identified as part of inventory on hand. They are treated as an asset until the goods they are assigned to are sold. At that time they become the expense, cost of goods sold. All manufacturing costs are product costs.

Product ———————→ Asset ——————→ Expense
  costs   (inventory)   (cost of goods sold)

Period costs are all expired costs that are not necessary for production and hence are charged against sales revenues in the period in which the revenue is earned. Selling and general and administrative expenses are period costs.

Product costs ————————————→ Expense

Variable Costs, Fixed Costs, and Mixed Costs

From a planning and control standpoint, perhaps the most important way to classify costs is by how they behave in accordance with changes in volume or some measure of activity. By behavior, costs can be classified into the following three basic categories: variable, fixed, and mixed costs.

Variable costs are costs that vary in total in direct proportion to changes in activity. Examples are direct materials and gasoline expense based on mileage driven. Fixed costs are costs that remain constant in total regardless of changes in activity. Examples are rent, insurance, and taxes. Mixed (or semi-variable) costs are costs that vary with changes in volume but, unlike variable costs, do not vary in direct proportion. In other words, these costs contain both a variable component and a fixed component. Examples are the rental of a delivery truck, where a fixed rental fee plus a variable charge based on mileage is made; and power costs, where the expense consists of a fixed amount plus a variable charge based on consumption.

The breakdown of costs into their variable components and their fixed components is important in many areas of management accounting, such as flexible budgeting, break-even analysis, and short-term decision making.

Unit Costs and Total Costs

For external reporting and pricing purposes, project managers are frequently interested in determining the unit (average) cost per unit of product or service. The unit cost is simply the average cost, which is the total costs divided by the

Example 5.1. Fixed costs are $1,000 per period and variable costs are $.10 per unit. The total and unit (average) costs at various production levels are as shown in the table.

Volume in units	Total fixed costs	Total variable costs	Total costs	Variable cost per unit	Fixed cost per unit	Unit (average) cost
(a)	(b)	(c)	(b)+(c) = (d)	(c)/(a) = (e)	(b)/(a) = (f)	(d)/(a) or (e)+(f)
1,000	$1,000	$ 100	$1,100	$.10	$1.00	$1.10
5,000	1,000	500	1,500	.10	.20	.30
10,000	1,000	1000	2,000	.10	.10	.20

total volume in units. Alternatively, the unit cost is the sum of (a) the variable cost per unit, and (b) the fixed cost per unit. It is important to realize that the unit cost declines as volume increases since the total fixed costs that are constant over a range of activity are being spread over a larger number of units.

The increase in total costs and the decline in unit costs are illustrated in Figure 5.2. Also note the relationships for variable and fixed costs per unit as volume changes from 5,000 to 10,000:

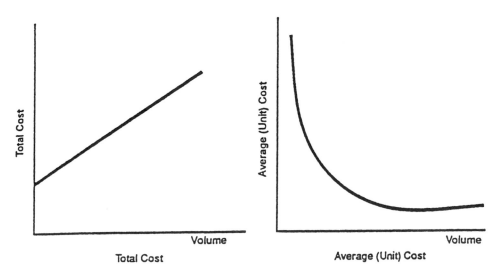

Figure 5.2. Total and unit (average) costs.

	Total Cost	Unit Cost
Variable cost	Change ($500 to $1,000)	No change ($.10)
Fixed cost	No change ($1,000)	Change ($.20 to $.10)

Costs for Planning, Control, and Decision Making

Controllable and Noncontrollable Costs

A cost is said to be controllable when the amount of the cost is assigned to the head of a department and the level of the cost is significantly under the project manager's influence. Noncontrollable costs are those costs not subject to influence at a given level of the project manager's supervision.

Example 5.2. All variable costs such as direct materials, direct labor, and variable overhead are usually considered controllable by the project manager. On the other hand, fixed costs such as depreciation of factory equipment would not be controllable by the project manager, since he/she would have no power to authorize the purchase of the equipment.

Standard Costs

Standard costs are the costs established in advance to serve as goals, norms or yardsticks to be achieved and, after the fact, to determine how well those goals were met. They are based on the quantities and prices of the various inputs (e.g., direct materials, direct labor, and factory overhead) needed to produce output efficiently. Standard costs can also be set for service businesses.

Example 5.3. The standard cost of materials per pound is obtained by multiplying standard price per pound by standard quantity per unit of output in pounds. For example, the standard price and quantity of material might be determined as follows:

Purchase price	$3.00
Freight	0.12
Receiving and handling	0.02
Less: Purchase discounts	(0.04)
Standard price per pound	$3.10
Per bill of materials in pounds	1.2
Allowance for waste and spoilage in lbs.	0.1
Allowance for rejects in lbs.	0.1
Standard quantity per unit of output	1.4 lbs.

Once the price and quantity standards have been set, the standard cost of material per unit of finished goods can be computed, as follows:

1.4 pounds × $3.10 = $4.34 per unit.

Incremental (or Differential) Costs

Incremental cost refers to differences in costs among alternatives. Incremental costs are increases or decreases in total costs or changes in specific elements of cost (e.g., direct labor cost), that result from any variation in operations.

Example 5.4. Consider the case of a manufacturer deciding whether to continue to make a particular part or to purchase the part from an outside supplier. The costs *relevant* to this decision are those which would actually be reduced or saved by discontinuing production and those which would actually be incurred by purchasing from an outside supplier. This topic will be explored more fully in Chapter 12.

Sunk Costs

Sunk costs are the costs of resources that have already been incurred whose total will not be affected by any decision made now or in the future. Sunk costs are considered *irrelevant* to future decisions since they are past or historical costs.

Example 5.5. Suppose you acquired an asset for $50,000 three years ago which is now listed at a book value of $20,000. The asset is worth $15,000. Neither the original purchase price nor the $20,000 book value is a sunk cost which does not affect a future decision. The relevant value is $15,000, the amount you could realize by selling it.

Out-of-Pocket Costs

Out-of-pocket costs, also known as outlay costs, are costs that require future expenditures of cash or other resources. Non-cash charges such as depreciation and amortization are not out-of-pocket costs. Out-of-pocket costs are usually relevant to a particular decision.

Example 5.6. A capital investment project requires $120,000 in cash outlays. $120,000 is an out-of-pocket cost.

Relevant Costs

Relevant costs are expected future costs that will differ between alternatives. This concept is a key to short- and long-term decisions and discussed in detail in Chapter 12.

Example 5.7. The incremental cost is said to be relevant to the future decision. The sunk cost is considered irrelevant.

Opportunity Costs

An opportunity cost is the net benefit foregone by rejecting an alternative. There is always an opportunity cost involved in making a choice decision. It is a cost incurred relative to the alternative given up.

Example 5.8. Suppose a company has a choice of using its capacity to produce an extra 10,000 units or renting it out for $20,000. The opportunity cost of using the capacity is $20,000.

INCOME STATEMENTS AND BALANCE SHEETS—MANUFACTURER

Figure 5.3 illustrates the income statement of a manufacturer. An important characteristic of the income statement is that it is supported by a schedule of cost of goods manufactured (Figure 5.4). This schedule shows the specific costs (i.e., direct materials, direct labor, and factory overhead) that have gone into the goods completed during the period. Since the manufacturer carries three types of inventory (raw materials, work-in-process, and finished goods) all three items must be incorporated into the computation of the cost of goods sold. These inventory accounts also appear on the balance sheet for a manufacturer, as shown in Figure 5.5.

Manufacturer's Income Statement
For the Year Ended December 31, 19x1

Sales		$320,000
Less: Cost of goods sold		
Finished goods, Dec.31 19x0	$ 18,000	
Cost of goods manufactured		
(see schedule, Figure 5.5)	121,000	
Cost of goods available		
for sale	$139,000	
Finished goods, Dec.31 19x1	21,000	
Cost of goods sold		$118,000
Gross margin		$202,000
Less: Selling and admini-		
strative expenses		60,000
Net income		$142,000

Figure 5.3. Manufacturer's income statement.

Manufacturer's Schedule of Cost Goods Manufactured

Direct Materials:		
Inventory, Dec. 31, 19x0	$23,000	
Purchases	64,000	
Cost of direct materials		
Available for use	$87,000	
Inventory, Dec. 31, 19x1	7,800	
Direct materials used		$ 79,200
Direct labor		25,000
Factory overhead:		
Indirect labor	$ 3,000	
Indirect material	2,000	
Factory utilities	500	
Factory depreciation	800	
Factory rent	2,000	
Miscellaneous	1,500	9,800
Total manufacturing costs		
Incurred during 19x1		$114,000
Add:Work-in-process inventory,		
Dec. 31, 19x0		9,000
Manufacturing costs to account for		$123,000
Less:Work-in-process inventory,		
Dec. 31, 19x1		2,000
Cost of goods manufactured (to		
income statement, Figure 5.3)		$121,000

Figure 5.4. Manufacturer's schedule of cost goods manufactured.

Manufacturer's Current Asset Section of Balance Sheet
December 31, 19x1

Current Assets:		
Cash		$ 25,000
Accounts receivable		78,000
Inventories:		
Raw materials	$ 7,800	
Work-in-process	2,000	
Finished goods	21,000	30,800
Total		$133,800

Figure 5.5. Manufacturer's current asset section of balance sheet.

ANALYSIS OF COST BEHAVIOR AND FLEXIBLE BUDGETING

For forecasting, planning, control, and decision making purposes, mixed costs need to be separated into their variable and fixed components. Since the mixed costs contain both fixed and variable elements, the analysis takes the following mathematical form, which is called a *cost-volume formula*:

$$Y = a + bX$$

where

> Y = the mixed cost to be broken up
> X = any given measure of activity such as direct labor hours, machine hours, or production volume
> a = the fixed cost component
> b = the variable rate per unit of X

Note: The formula is also called the *flexible budget formula*, since it is used to generate budgeted (estimated) costs for various levels of activity.

Separating the mixed cost into its fixed and variable components is the same thing as estimating the parameter values a and b in the cost-volume formula. There are several methods available to be used for this purpose including the high-low method and regression analysis.

THE HIGH-LOW METHOD

The high-low method, as the name indicates, uses two extreme data points to determine the values of a (the fixed cost portion) and b (the variable rate) in the equation $Y = a + bX$. The extreme data points are the highest representative X—Y pair and the lowest representative X—Y pair. The activity level X, rather than the mixed cost item Y, governs their selection.

The high-low method is explained, step by step, as follows:

Step 1: Select the highest pair and the lowest pair
Step 2: Compute the variable rate, b, using the formula:

$$\text{Variable rate} = \frac{\text{Difference in cost Y}}{\text{Difference in activity X}}$$

Step 3: Compute the fixed cost portion as:

Fixed cost portion = Total mixed cost − Variable cost

Table 5.1

Month	Direct Labor Hours (X)	Factory Overhead (Y)
April	2750	99
May	2330	93
June	2690	103
July	2480	77
August	2610	102
September	2910	122

Example 5.9. Flexible Manufacturing Company decided to relate total factory overhead costs to direct labor hours (DLH) to develop a cost-volume formula in the form of $Y = a + b X$. Twelve monthly observations are collected. They are given in Table 5.1.

The high-low points selected from the monthly observations are

	X	Y
High	122 hours	$2920 (September pair)
Low	93	2330 (May pair)
Difference	29 hours	$ 590

Thus

$$\text{Variable rate } b = \frac{\text{Difference in Y}}{\text{Difference in X}} = \frac{\$590}{29 \text{ hours}} = \$20.34 \text{ per DLH}$$

The fixed cost portion is computed as

$$
\begin{aligned}
\text{Fixed cost portion} &= \text{Total mixed cost - Variable cost} \\
&= \$2920 - (\$20.34)(122 \text{ hours}) \\
&= \$2920 - \$2481.48 = \$438.52
\end{aligned}
$$

Therefore, the cost formula for factory overhead is $438.52 fixed plus $20.34 per DLH.

The high-low method is simple and easy to use. It has the disadvantage, however, of using two extreme data points, which may not be representative of normal conditions. The method may yield unreliable estimates of a and b in our formula. In this example, the negative value for a is questionable. In such a case, it would be wise to drop them and choose two other points that are more representative of normal situations.

The Least-Squares Method

One popular method for estimating the cost-volume formula is *the least-squares method* (or regression analysis). Unlike the high-low method, in an effort to estimate the variable rate and the fixed cost portion, this method includes all the observed data and attempts to find a line of best fit. The method is reserved for an advanced cost accounting text.

FLEXIBLE BUDGETING

A flexible budget is a tool that is extremely useful in cost control. In contrast to a static budget, the flexible budget is characterized as follows:

1. It is geared toward a range of activity rather than a single level of activity.
2. It is dynamic in nature rather than static. By using the cost-volume formula (or flexible budget formula), a series of budgets can be easily developed for various levels of activity.

The primary use of the flexible budget is to accurately measure performance by comparing actual costs for a given output with the budgeted costs for the same level of output. This topic will be discussed in detail in Chapter 8.

ACCUMULATION OF COSTS AND COST ACCOUNTING

A cost accumulation system is a product costing system. This process accumulates manufacturing costs such as materials, labor and factory overhead and assigns them to cost objectives, such as finished goods and work-in-process. Product costing is necessary not only for inventory valuation and income determination but also for establishing the unit sales price.

We will discuss the essentials of the cost accumulation system that is used to measure the manufacturing costs of products. This is essentially a two-step process: (1) the measurement of costs that are applicable to manufacturing operations during a given accounting period, and (2) the assignment of these costs to products.

There are two basic approaches to cost accounting and accumulation:

1. Job order costing
2. Process costing

		Job order costing	Process costing
1.	Cost unit	Job, order, or contract	Physical unit
2.	Costs are accumulated	By jobs	By departments
3.	Subsidiary record	Job cost sheet	Cost of production report
4.	Used by	Custom manufacturers	Processing industries
5.	Permits computation of:	a. A unit cost for inventory costing purposes	A unit cost to be used to compute the costs of goods completed and work in process
		b. A profit or loss on each job	

Figure 5.6. Differences between job order costing and process costing.

JOB ORDER COSTING AND PROCESS COSTING COMPARED

The distinction between job order costing and process costing centers largely around how product costing is accomplished. With job order costing, the focus is to apply costs to specific jobs, which may consist of either a single physical unit or a few like units.

Under process costing, data are accumulated by the production department (or cost center) and averaged over all of the production that occurred in the department. Here there is mass production of like units which are manufactured on a continuous basis through a series of uniform production steps known as *processes*. Figure 5.6 summarizes the basic differences between these two methods.

JOB ORDER COSTING

Job order costing is the cost accumulation system under which costs are accumulated by jobs, contracts, or orders. This costing method is appropriate when the products are manufactured in identifiable lots or batches or when the products are manufactured to customer specifications. Job order costing is widely used by custom manufacturers such as printing, aircraft, and construction companies. It may also be used by service businesses such as auto repair shops and professional services. Job order costing keeps track of costs as follows: direct material and direct labor are traced to a particular job. Costs not directly traceable—factory overhead—are applied to individual jobs using a predetermined overhead (application) rate.

Example 5.10. Holden Works collects its cost data by the job order cost system. For Job 123, the following data are available:

Direct materials		Direct labor
7/14 Issued	$1,200	Week of July 20 180 hrs. @$6.50
7/20 Issued	650	Week of July 26 140 hrs. @ 7.25
7/25 Issued	350	
	$2,200	

Factory overhead is applied at the rate of $4.50 per direct labor hour.

We will compute (1) the cost of Job 123 and (2) the sales price of the job, assuming that it was contracted with a markup of 40% of cost.

(1) The cost of the job is:

Direct material		$2,200
Direct labor:		
180 hrs. × $6.50	$1,170	
140 hrs. × $7.25	1,015	2,185
Factory overhead applied:		
320 hrs. × $4.50		1,440
Cost of job 123		$5,825

(2) The sales price of the job is:

$$\$5,825 + 40\% \ (\$5,825) = \$5,825 + \$2,330 = \$8,155$$

PROCESS COSTING

Process costing is a cost-accumulation system that aggregates manufacturing costs by departments or by production processes. Total manufacturing costs are accumulated by two major categories, direct materials and conversion costs (the sum of direct labor and factory overhead applied). Unit cost is determined by dividing the total costs charged to a cost center by the output of that cost center. In this sense, the unit costs are averages.

Process costing is appropriate for companies that produce a continuous mass of like units through a series of operations or processes. Process costing is generally used in such industries as petroleum, chemicals, oil refinery, textiles, and food processing.

Steps in Process Costing Calculations

There are basically five steps to be followed in accounting for process costs.

1. Summarize the flow of physical units.
The first step of the accounting provides a summary of all units on which some work was done in the department during the period. *Input must equal output.* This step helps to detect lost units during the process. The basic relationship may be expressed in the following equation:

Beginning inventory + Units started for the period =
    Units completed and transferred out + Ending inventory

2. Compute output in terms of equivalent units.
In order to determine the unit costs of the product in a processing environment, it is important to measure the total amount of work done during an accounting period. A special problem arises in processing industries in connection with how to deal with work still in process, that is, the work partially completed at the end of the period. The partially completed units are measured on an equivalent whole-unit basis for process costing purposes.

Equivalent units are a measure of how many whole units of production are represented by the units completed plus the units partially completed. For example, 100 units that are 60% completed are the equivalent of 60 completed units.

3. Summarize the total costs to be accounted for by cost categories.
This step summarizes the total costs assigned to the department during the period.

4. Compute the unit costs per equivalent unit.
The unit costs per equivalent is computed as follows:

$$\text{Unit cost} = \frac{\text{Total costs incurred during period}}{\text{Equivalent production units during period}}$$

5. Apply total costs to units completed and transferred out and to units in ending work-in-process.

Cost-of-Production Report

The process costing method uses what is called the cost-of-production report. It summarizes both total and unit costs charged to a department and indicates the allocation of total costs between work-in-process inventory and the units completed and transferred out to the next department or the finished goods inventory.

The cost-of-production report covers all five steps described above. It is a convenient compilation from which cost data may be presented to management. The following example illustrates the report.

Example 5.11. (Process Cost Computation—No Beginning Inventory) A company produces and sells a chemical product that is processed in two departments. In Department A, the basic materials are crushed, powdered, and mixed. In Department B, the product is tested, packaged, and labeled, and before being transferred to finished goods inventory.

Assume the following for Production Department A for May. Materials are added when production is begun; therefore, all finished units and all units in the ending work-in-process inventory will have received a full complement of materials.

Actual production costs:
  Direct materials used, 18,000 gallons costing $27,000
  Direct labor and factory overhead, $25,000
Actual production:
  Completed and transferred to Production Department B, 8,000 gallons
  Ending work-in-process, 10,000 gallons, 20% complete as to conversion

1. *Summarize the flow of physical units.*

To be accounted for:
  Added this period   18,000 gallons

Accounted for as follows:
  Completed this period   8,000 gallons
  In process, end of period 10,000
    Total   18,000 gallons

2. *Compute output in terms of equivalent units.*

	Materials (gal.)	Conversion cost
Units completed	8,000	8,000
Ending work-in-process (10,000 gallons)		
100% of materials	10,000	
20% of conversion cost		2,000
Equivalent units produced	18,000	10,000

Steps 3. through 5.

These steps are summarized in the Cost of Production report.

Cost of Production

	Total cost	Equivalent production (gal.)	Unit cost
Materials	$27,000	18,000	$1.50
Conversion cost	25,000	10,000	2.50
To be accounted for	$52,000		$4.00
Ending work-in-process:			
Materials	$15,000	10,000	$1.50
Conversion cost	5,000	2,000	2.50
Total work-in-process	$20,000		
Completed and transferred	32,000	8,000	4.00
Total accounted for	$52,000		

CONCLUSION

Cost accounting is the accumulation and analysis of cost data to provide information for external reporting, for internal planning and control of an organization's operations, and for short-term and long-term decisions. It is important to realize that there are different costs used for different purposes. The engineer needs to have an understanding of how to use cost data. He/she must investigate cost behavior for break-even, for appraisal of managerial performance, and for flexible budgeting. We have looked at three types of cost behavior—variable, fixed, and mixed. We discussed two popular methods of separating mixed costs in their variable and fixed components: the high-low and least-squares methods. The idea of flexible budgeting was emphasized in an attempt to correctly measure the efficiency of the manufacturing department.

Unit costs are necessary for inventory valuation, income determination, and pricing. This chapter provided an introduction to the two basic cost accumulation systems: (1) job order costing, and (2) process costing.

Job order costing attaches costs to specific jobs by means of cost sheets established for each job. Direct material and direct labor costs are traced to specific jobs and factory overhead costs are applied by jobs, using a predetermined overhead rate.

Process costing makes no attempt to cost any specific lot in process. All costs, direct and indirect, are accumulated by departments for periods of time and an average cost for the period is computed. The five steps in process costing determination are: the check of a physical flow, computation of equivalent units, summary of costs, calculation of the unit costs, and calculation of the cost of goods completed and the ending work-in-process. The choice of the system, job order or process, depends on the nature of the manufacturing operation and the desired information.

6

BUDGETING FOR PROFIT PLANNING

A comprehensive (master) budget is a formal statement of management's expectations regarding sales, expenses, volume, and other financial transactions of an organization for the coming period. Simply put, a budget is a set of *pro forma* (projected or planned) financial statements. It consists basically of a pro forma income statement, pro forma balance sheet and cash budget.

A budget is a tool for both planning and control. At the beginning of the period, the budget is a plan or standard; at the end of the period it serves as a control device to help management measure its performance against the plan so that future performance may be improved.

It is important to realize that with the aid of computer technology, budgeting can be used as an effective device for evaluation of "what-if" scenarios. This way management should be able to move toward finding the best course of action among various alternatives through simulation.

If management does not like what they see on the budgeted financial statements in terms of various financial ratios such as liquidity, activity (turnover), leverage, profit margin, and market value ratios, they can always alter their contemplated decision and planning set.

TYPES OF BUDGETS

The budget is classified broadly into two categories:

1. Operating budget, reflecting the results of operating decisions.
2. Financial budget, reflecting the financial decisions of the firm.

The operating budget consists of:

Sales budget
Production budget
Direct materials budget
Direct labor budget
Factory overhead budget
Selling and administrative expense budget
Pro forma income statement

The financial budget consists of:

Cash budget
Pro forma balance sheet

The major steps in preparing the budget are:

1. Prepare a sales forecast.
2. Determine expected production volume.
3. Estimate manufacturing costs and operating expenses.
4. Determine cash flow and other financial effects.
5. Formulate projected financial statements.

Figure 6.1 shows a simplified diagram of the various parts of the comprehensive (master) budget, the master plan of the company.

ILLUSTRATION

To illustrate how all these budgets are put together, we will consider a manufacturing company called the Pelican Company, which produces and markets a single product. We will assume that the company develops the master budget in contribution format for 19x1 on a quarterly basis. We will highlight the variable cost-fixed cost breakdown throughout the illustration.

The Sales Budget

The sales budget is the starting point in preparing the master budget, since estimated sales volume influences nearly all other items appearing throughout the

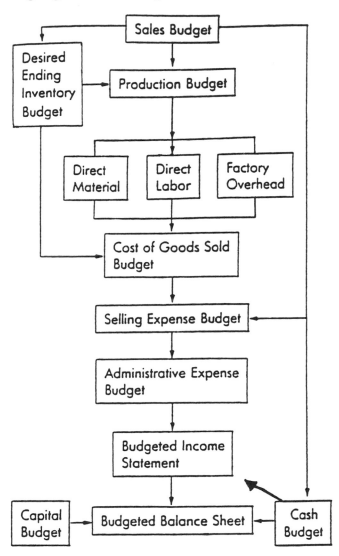

Figure 6.1. Comprehensive (master) budget.

master budget. The sales budget ordinarily indicates the quantity of each product expected to be sold. After sales volume has been estimated, the sales budget is constructed by multiplying the expected sales in units by the expected unit sales price. Generally, the sales budget includes a computation of expected cash collections from credit sales, which will be used later for cash budgeting.

Example 6.1. The Pelican Company's sales budget and schedule of expected cash collections follows:

The Pelican Company
Sales Budget
For the Year Ending December 31, 19x1

	Quarter				
	1	2	3	4	Total
Expected sales in units	800	700	900	800	3,200
Unit sales price	× $80	× $80	× $80	× $80	× $80
Total sales	$64,000	$56,000	$72,000	$64,000	$256,000

Schedule of Expected Cash Collections

Accounts receivable,12/31/19x1: $ 9,500[a]					$ 9,500
1st quarter sales ($64,000)	44,800[b]	$17,920[c]			62,720
2d quarter sales ($56,000)		39,200	$15,680		54,880
3d quarter sales ($72,000)			50,400	$20,160	70,560
4th quarter sales ($64,000)				44,800	44,800
Total cash collections	$54,300	$57,120	$66,080	$64,960	$242,460

[a]All $9,500 accounts receivable balance is assumed to be collectible in the first quarter.
[b]70% of a quarter's sales are collected in the quarter of sale.
[c]28% of a quarter's sales are collected in the quarter following, and the remaining 2% are uncollectible.

The Production Budget

After sales are budgeted, the production budget can be determined. The number of units expected to be manufactured to meet budgeted sales and inventory requirements is set forth in the production budget. The expected volume of production is determined by subtracting the estimated inventory at the beginning of the period from the sum of the units expected to be sold and the desired inventory at the end of the period. The production budget is illustrated on the following page.

The Direct Material Budget

When the level of production has been computed, a direct material budget should be constructed to show how much material will be required for production and how much material must be purchased to meet this production requirement.

Example 6.2. The Pelican Company's production budget follows:

The Pelican Company
Production Budget
For the Year Ending December 31, 19x1

	Quarter				
	1	2	3	4	Total
Planned sales (Ex.6.1)	800	700	900	800	3,200
Desired ending inventory[a]	70	90	80	100[b]	100
Total Needs	870	790	980	900	3,300
Less: Beginning inventory[c]	80	70	90	80	80
Units to be produced	790	720	890	820	3,220

[a]10% of the next quarter's sales.
[b]Estimated.
[c]The same as the previous quarter's ending inventory.

The purchase will depend on both expected usage of materials and inventory levels. The formula for computation of the purchase is:

Purchase in units = Usage + Desired ending material inventory units
− Beginning inventory units

The direct material budget is usually accompanied by a computation of expected cash payments for materials.

The Direct Labor Budget

The production requirements as set forth in the production budget also provide the starting point for the preparation of the direct labor budget. To compute direct labor requirements, expected production volume for each period is multiplied by the number of direct labor hours required to produce a single unit. The direct labor hours to meet production requirements is then multiplied by the (standard) direct labor cost per hour to obtain budgeted total direct labor costs.

The Factory Overhead Budget

The factory overhead budget should provide a schedule of all manufacturing costs other than direct materials and direct labor. Using the contribution

Example 6.3. The Pelican Company's direct material budget and schedule of expected cash disbursements follows:

The Pelican Company
Direct Material Budget
For the Year Ending December 31, 19x1

	Quarter				
	1	2	3	4	Total
Units to be produced (Ex. 6.2)	790	720	890	820	3,220
Material needs per unit(lbs)	× 3	× 3	× 3	× 3	× 3
Material needs for production	2,370	2,160	2,670	2,460	9,660
Desired ending inventory of materials[a]	216	267	246	250[b]	250
Total needs	2,586	2,427	2,916	2,710	9,910
Less: Beginning inventory of materials[c]	237	216	267	246	237
Materials to be purchased	2,349	2,211	2,649	2,464	9,673
Unit price	× $2	× $2	× $2	× $2	× $2
Purchase cost	$4,698	$4,422	$5,298	$4,928	$19,346

Schedule of Expected Cash Disbursements

	1	2	3	4	Total
Accounts payable, 12/31/19A	$2,200				$ 2,200
1st quarter purchases($4,698)	2,349	2,349[d]			4,698
2d quarter purchases ($4,422)		2,211	2,211		4,422
3d quarter purchases ($5,298)			2,649	2,649	5,298
4th quarter purchases ($4,928)				2,464	2,464
Total disbursements	$4,549	$4,560	$4,860	$5,113	$19,082

[a]10% of the next quarter's units needed for production.
[b]Estimated.
[c]The same as the prior quarter's ending inventory.
[d]50% of a quarter's purchases are paid for in the quarter of purchase; the remainder are paid for in the following quarter.

approach to budgeting requires the development of a predetermined overhead rate for the variable portion of the factory overhead. In developing the cash budget, we must remember that depreciation does not entail a cash outlay and therefore must be deducted from the total factory overhead in computing cash disbursement for factory overhead.

Example 6.4. The Pelican Company's direct labor budget follows:

The Pelican Company
Direct Labor Budget
For the Year Ending December 31, 19x1

	Quarter				
	1	2	3	4	Total
Units to be produced(Ex.6.2)	790	720	890	820	3,220
Direct labor hours per unit	× 5	× 5	× 5	× 5	× 5
Total hours	3,950	3,600	4,450	4,100	16,100
Direct labor cost per hour	× $5	× $5	× $5	× $5	× $5
Total direct labor cost	$19,750	$18,000	$22,250	$20,500	$80,500

Example 6.5. To illustrate the factory overhead budget, we will assume that:

Total factory overhead budgeted = $6,000 fixed(per quarter), plus $2 per hour of direct labor.
Depreciation expenses are $3,250 each quarter.
All overhead costs involving cash outlays are paid for in the quarter incurred.

The Pelican Company
Factory Overhead Budget
For the Year Ending December 31, 19x1

	Quarter				
	1	2	3	4	Total
Budgeted direct labor hours (Ex.6.4)	3,950	3,600	4,450	4,100	16,100
Variable overhead rate	× $2	× $2	× $2	× $2	× $2
Variable overhead budgeted	$ 7,900	$ 7,200	$ 8,900	$ 8,200	$32,200
Fixed overhead budgeted	6,000	6,000	6,000	6,000	24,000
Total budgeted overhead	$13,900	$13,200	$14,900	$14,200	$56,200
Less: Depreciation	3,250	3,250	3,250	3,250	13,000
Cash disbursement for overhead	$10,650	$ 9,950	$11,650	$10,950	$43,200

Example 6.6. The Pelican Company's year-end inventory budget follows:

The Pelican Company
Ending Inventory Budget
For the Year Ending December 31, 19x1

	Ending Inventory		
	Units	Unit cost	Total
Direct materials	250 pounds (Ex. 3)	$2	$ 500
Finished goods	100 units (Ex. 2)	$41[a]	$4,100

[a]The unit variable cost of $41 is computed as follows:

	Unit cost	Units	Total
Direct materials	$2	3 pounds	$ 6
Direct labor	5	5 hours	25
Variable overhead	2	5 hours	10
Total variable manufacturing cost			$41

The Ending Inventory Budget

The desired ending inventory budget provides us with the information required for the construction of budgeted financial statements. Specifically, it will help compute the cost of goods sold on the budgeted income statement. Secondly, it will give the dollar value of the ending materials and finished goods inventory to appear on the budgeted balance sheet.

The Selling and Administrative Expense Budget

The selling and administrative expense budget lists the operating expenses involved in selling the products and in managing the business. In order to complete the budgeted income statement in contribution format, variable selling and administrative expense per unit must be computed.

The Cash Budget

The cash budget is prepared for the purpose of cash planning and control. It presents the expected cash inflow and outflow for a designated time period. The cash budget helps management keep cash balances in reasonable relationship to its needs. It aids in avoiding unnecessary idle cash and possible cash shortages. The cash budget consists typically of four major sections:

Example 6.7. The Pelican Company's selling and administrative expense budget
follows:

The Pelican Company
Selling and Administrative Expense Budget
For the Year Ending December 31, 19x1

	Quarter				
	1	2	3	4	Total
Expected sales in units	800	700	900	800	3,200
Variable selling and admin. exp. per unit[a]	× $4	× $4	× $4	× $4	× $4
Budgeted variable expense	$ 3,200	$ 2,800	$ 3,600	$ 3,200	$12,800
Fixed selling and administrative expenses:					
Advertising	1,100	1,100	1,100	1,100	4,400
Insurance	2,800				2,800
Office salaries	8,500	8,500	8,500	8,500	34,000
Rent	350	350	350	350	1,400
Taxes			1,200		1,200
Total budgeted selling and administrative expenses[b]	$15,950	$12,750	$14,750	$13,150	$56,600

[a]Includes sales agents' commissions, shipping, and supplies.
[b]Paid for in the quarter incurred.

1. The *receipts* section, which is the beginning cash balance, cash collections from customers, and other receipts
2. The *disbursements* section, which comprises all cash payments made by
   purpose
3. The *cash surplus or deficit* section, which simply shows the difference
   between the cash receipts section and the cash disbursements section
4. The *financing* section, which provides a detailed account of the borrowings and repayments expected during the budgeting period

Example 6.8. To illustrate, we will make the following assumptions:

The company desires to maintain a $5,000 minimum cash balance at the end
   of each quarter.
All borrowing and repayment must be in multiples of $500 at an interest rate
   of 10% per annum.
Interest is computed and paid as the principal is repaid.
Borrowing takes place at the beginning of each quarter and repayment at the
   end of each quarter.

The Pelican Company
Cash Budget
For the Year Ending December 31, 19x1

		Quarter				
	Example	1	2	3	4	Total
Cash balance, beginning	Given	$10,000	$ 9,401	$ 5,461	$ 9,106	$ 10,000
Add: Receipts:						
Collection from customers	6.1	54,300	57,120	66,080	64,960	242,460
Total cash available		$64,300	$66,521	$71,541	$74,066	$252,460
Less: Disbursements:						
Direct materials	6.3	4,549	4,560	4,860	5,113	19,082
Direct labor	6.4	19,750	18,000	22,250	20,500	80,500
Factory overhead	6.5	10,650	9,950	11,650	10,950	43,200
Selling and Admin.	6.6	15,950	12,750	14,750	13,150	56,600
Machinery purchase	Given	—	24,300	—	—	24,300
Income tax	Given	4,000	—	—	—	4,000
Total disbursements		$54,899	$69,560	$53,510	$49,713	$227,682
Cash surplus (deficit)		$ 9,401	$ (3,039)	$18,031	$24,353	$ 24,778
Financing:						
Borrowing		—	8,500	—	—	8,500
Repayment		—	—	(8,500)	—	(8,500)
Interest		—	—	(425)	—	(425)
Total financing		—	$ 8,500	$(8,925)	—	$ (425)
Cash balance, ending		$ 9,401	$ 5,461	$ 9,106	$24,353	$ 24,353

The Budgeted Income Statement

The budgeted income statement summarizes the various component projections of revenue and expenses for the budgeting period. However, for control purposes, the budget can be divided into quarters or even months depending on the need.

The Budgeted Balance Sheet

The budgeted balance sheet is developed by beginning with the balance sheet for the year just ended and adjusting it, using all the activities that are expected to

Example 6.9. The Pelican Company's budgeted income statement is shown.

The Pelican Company
Budgeted Income Statement
For the Year Ending December 31, 19x1

	From Example		
Sales (3,200 units @ $80)	6.1		$256,000
Less: Variable expenses			
Variable cost of goods sold			
(3,200 units @ $41)	6.6	$131,200	
Variable selling and admin.	6.7	12,800	144,000
Contribution margin			$112,000
Less: Fixed expenses			
Factory overhead	6.5	$ 24,000	
Selling and admin.	6.7	43,800	67,800
Net operating income			$ 44,200
Less: Interest expense	6.8		425
Net income before taxes			$ 43,775
Less: Income taxes	20%		8,755
Net income			$ 35,020

take place during the budgeting period. Some of the reasons why the budgeted balance sheet must be prepared are:

It could disclose some unfavorable financial conditions that management might want to avoid.

It serves as a final check on the mathematical accuracy of all the other schedules.

It helps management perform a variety of ratio calculations.

It highlights future resources and obligations.

Some Financial Calculations

To see what kind of financial condition the Pelican Company is expected to be in for the budgeting year, a sample of financial ratio calculations are in order: (Assume 19x1 after-tax net income was $15,000.)

Example 6.10. To illustrate, we will use the following balance sheet for the year 19x1.

The Pelican Company
Balance Sheet
For the Year Ending December 31, 19x1

Assets		Liabilities and Stock Equity	
Current assets:		Current liabilities:	
Cash	$ 10,000	Accounts payable	$ 2,200
A/R	9,500	Income tax payable	4,000
Material inv.	474	Total cur. liab.	$ 6,200
Finished gd inv.	3,280		
	$ 23,254		
Fixed assets:		Stockholders' equity:	
Land	50,000	Common stock, no-par	70,000
Build and Eqpt	100,000	Retained earnings	37,054
Accumtd Depr	(60,000)		
	$ 90,000		
Total assets	$113,254	Total liab. and stk. eq.	$113,254

The Pelican Company
Budgeted Balance Sheet
For the Year Ending December 31, 19x1

Assets		Liabilities and Stock Equity	
Current assets:		Current liabilities:	
Cash	24,353[a]	Accounts payable	2,464[h]
Accounts rec	23,040[b]	Income tax payable	8,755[i]
Material inv	500[c]	Total cur. liab.	$ 11,219
Finished gd. inv	4,100[d]		
	$ 51,993		
Fixed assets:		Stockholders' equity:	
Land	50,000[e]	Common stock, no-par	70,000[j]
Build. and eqpt.	124,300[f]	Retained earnings	72,074[k]
Accumtd depr.	(73,000)[g]		
	$101,300		
Total assets	$153,293	Total liab. and stk. eq.	$153,293

	19x1	19x0
Current ratio		
(Current assets/	$23,254/$6,200	$51,993/$11,219
current liabilities)	= 3.75	= 4.63
Return on total assets		
(Net income after taxes/	$15,000/$113,254	$35,020/$153,293
total assets)	= 13.24%	= 22.85%

Computations:
[a]From Example 6.8 (cash budget).
[b]$9,500 + $256,000 sales − $242,460 receipts = $23,040.
[c] and [d] from Example 6.6 (ending inventory budget).
[e]No change.
[f]$100,000 + $24,300 (from Example 6.8) = $124,300.
[g]$60,000 + $13,000 (from Example 6.5) = $73,000.
[h]$2,200 + $19,346 − $19,082 = $2,464 (all accounts payable relate to material purchases),
or 50% of 4th quarter purchase = 50% ($4,928) = 2,464.
[i]From Example 6.9 (budgeted income statement).
[j]No change.
[k]$37,054 + $35,020 net income = $72,074.

Sample calculations indicate that the Pelican Company is expected to have better liquidity as measured by the current ratio. Overall performance will be improved as measured by return on total assets. This could be an indication that the contemplated plan may work out well.

A SHORT-CUT APPROACH TO FORMULATING THE BUDGET

In actual practice, use of a shortcut approach is widely used in formulating a budget. The approach can be summarized as follows:

1. A pro forma income statement is developed using past percentage relationships between certain expense and cost items and the firm's sales and applying these percentages to the firm's projected sales. This can be done using a spreadsheet program such as Lotus 1-2-3.
2. A pro forma balance sheet is estimated, using the precentage-of-sales-method, which involves the following steps:
   a. Express balance sheet items that vary directly with sales as a percentage of sales. Any item that does not vary with sales (such as long-term debt) is designated not applicable (n.a.). Multiply these percentages by the sales projected to obtain the amounts for the future period.

    b. Where no percentage applies (such as long-term debt, common stock, and paid-in-capital), simply insert the figures from the present balance sheet or their "desired" level in the column for the future period.

    c. Compute the projected retained earnings as follows: Projected retained earnings = Present retained earnings + Projected net income − Cash dividend to be paid

    d. Sum the asset accounts and the liability and equity accounts to see if there is any difference. The difference, if any, is a shortfall, which is the amount of financing the firm has to raise externally.

Computer-Based Models and Spreadsheet Program Models for Budgeting

Besides using spreadsheet programs such as Lotus 1-2-3, more and more companies are developing computer-based models for financial planning and budgeting, using powerful, yet easy-to-use, financial modeling languages such as Comshare's *Interactive Financial Planning System (IFPS)* and Social Systems' *SIMPLAN*. These models help not only to build a budget for profit planning but answer a variety of "what-if" scenarios. The resultant calculations provide a basis for choice among alternatives under conditions of uncertainty. As indicated earlier, financial modeling can also be accomplished using spreadsheet programs such as *Lotus 1-2-3* and *Microsoft's Excel*.

ZERO-BASE BUDGETING

The traditional budgeting techniques involve adding or subtracting a given percentage increase or decrease to the preceding period's budget and arriving at a new budget. The prior period's costs are considered to be basic and the emphasis is usually placed on what upward revisions are to be made for the upcoming year. The traditional method focuses on inputs rather than outputs related to goal achievement and as such never calls for the evaluation of corporate activities from a cost/benefit perspective.

    Zero-Base Budgeting (ZBB) can generally be described as a technique which requires each manager to justify his entire budget request in detail from a base of zero and as such asks for an analysis of the output values of each activity of a particular cost/responsibility center. This approach requires that all activities under scrutiny be defined in decision packages which are to be evaluated and ranked in order of importance at various levels. As an end product, a body of structured data is obtained that enables management to allocate funds confidently to the areas of greatest potential gain.

    ZBB is most applicable in planning service and support expenses rather than direct manufacturing expenses. This technique is best suited to operations and programs over which management has some discretion. For example, it can be used to develop:

Administrative and general support
Marketing
Research
Engineering
Manufacturing support
Capital budgets

It should not be used for:

Direct labor
Direct material
Factory overhead

which are usually budgeted through various methods discussed in the previous section. The following companion list helps our understanding of ZBB by indicating the key differences between ZBB and traditional (incremental) budgeting systems.

Traditional	Zero-base
Starts from existing base	Starts with base zero
Examines cost/benefit for new activities	Examines cost/benefit for all activities
Starts with dollars	Starts with purposes and activities
Does not examine new ways of operating as integral part of process	Explicitly examines new approaches
Results in a non-alternative budget	Results in a choice of several levels of service and cost

Methodology

ZBB is designed to result in a more rational and efficient allocation of firm resources and replanning support functions by requiring the evaluation of segment overhead activity relative to desired output (company goals). The various actions one would have to take in implementing ZBB are outlined below.

1. Setting Objectives and Assumptions

The business objectives and plan assumptions begin the zero-base budgeting process. Plan assumptions serve as input to the various operating departments in preparing their individual budgets. To effectively analyze the operation, lower management will need planning assumptions about inflation rates, salary increases, etc. It is possible that in beginning phases, a company would want to test ZBB in a specific division before widespread organizational application.

2. Defining Identifiable Segments

An Activity Unit is the basic cost element which is the subject of ZBB. An Activity Unit is usually made up of a group of employees who work toward a common goal. Such Activity Units are normally broken down into the traditional boundaries of a business although ZBB allows further definition. Activity Units or Division Units can be analyzed for discretionary activity. Activities fixed by law, industry practice, or other constraints are distinguished from those where action can be effected.

These decision units need to be established at an organizational level high enough so that the person responsible for the unit has effective control over the activities. Furthermore, it is desirable for decision units to be roughly similar in size (in terms of personnel and dollars) to allow effective comparison and the definition of the activity units should be specific enough to avoid complications arising from including a multiplicity of activities in a decision unit.

3. Decision Unit Analysis

1. Description of Current Practice

Following the listing of activity objectives, the decision manager describes how his department currently operates and the resources used (people/dollars). Operation description cannot become overly detailed but should contain the essentials of activities performed and are usually organized to define the flow of work.

2. Work Load and Performance Measurement

Performance measurements are next developed to examine the productivity and effectiveness of the manager's current approach. Some sample performance measurements might include:

Production control	On time delivery performance
Regional sales manager	Number of customer requests for cancellations
Internal audit	Cycle for audit coverage of reporting units
Quality control	Number of shipment deficiencies

3. Alternatives

ZBB next requires that the manager consider alternative ways of operating. After reviewing both the current and alternative operation methods, the manager and superiors will attempt to select the best method of operation on the basis of this analysis which will define the advantages and disadvantages of each method.

Examples of different operating modes would be:

Centralizing the function
Decentralizing the function
Contracting for the function
Combining the function with other activities
Eliminating the function

4. Ranking Analysis

In this step, the manager determines which is the most important service provided by his unit.

The highest priority is given to the minimum increment of service; the amount of service that the organization must undertake to provide any meaningful service. Additional increments are developed with each successive increment containing those services which are next in order of priority. Workload and performance measurements are included in each analysis since they identify meaningful quantitative measurements to assist in the activity unit decision process.

5. Review and Reallocate Resources

The increments developed by unit managers provide top-level management with the basic information for resource allocation decisions. The prioritization of service levels is the key factor in this process. Ranking takes place when the manager meets with all of the decision unit managers to prioritize the unit activities based on group objectives.

Ranking is based on discussions between the decision unit managers and the ranking manager and written cost-benefit analyses provided by each decision unit. Once ranking has been performed at various levels, a ranking table is prepared as a record of all the decisions that have been made in ZBB. This ranking table will indicate what will and will not be funded and rank activities in priorities to allow for easy adjustments to be made during the year.

6. Detailed Budgets Prepared

Once allocation decisions are made, detail budgets are prepared. These budgets usually are prepared on the basis of incremental activities indicated on management's ranking table.

7. Evaluate Performance

ZBB provides financial data as well as work load and performance measurements that can be monitored periodically. To be effective ZBB needs to be measured and controlled. Some control measures include the following:

Monthly financial review of each unit based on costs expended, actual vs.
   budget
Quarterly output review based on preestablished performance measurements
Quarterly plan and budget revisions for company and decision units based
   on to-date performance, changed environment factors.

Figure 6.2 offers a graphic illustration of the steps a company would take in
developing and implementing zero-base budgeting.

Advantages and Disadvantages

The advantages and disadvantages of ZBB can be defined as follows:

   1. *Advantages*

   a.   ZBB creates an analytical atmosphere which promotes the reorganiza-
        tion of activities to a more efficient mode and causes the evaluation of
        tasks from an output perspective.
   b.   ZBB involves line managers in the budgeting process and as such fos-
        ters support for implementation down through the company levels.
   c.   ZBB allows top management to define those service levels required
        from each business segment.
   d.   ZBB matches service levels to available resources and ensures over-
        heads are appropriate to the market place.

Senior Management			
	Planning assumptions	Ranking allocation	Evaluation and control
		Budget Preparation	
	Identify decision units	Decision unit analysis	
Middle Management			

Figure 6.2. Zero-Base Budgeting process.

2. *Disadvantages*

a. ZBB is perceived as an implied threat to existing programs.
b. ZBB requires a good data system to support analysis and in many cases no such system exists.
c. ZBB increases the demand of time placed on line managers.
d. Managers tend to overlook the goal of ZBB in evaluating activity units and focus on personnel security and interests.
e. Thrust usually comes from top to bottom and subordinates see little benefit for themselves.
f. The thought of creating a budget from scratch usually causes considerable resistance given the lack of support groups and training programs.

CONCLUSION

A budget is a detailed, quantitative plan outlining the acquisition and use of financial and other resources of an organization over some given time period. It is a tool for planning. If properly constructed, it is used as a control device. This chapter showed, step-by-step, how to formulate a master budget. The process begins with the development of a sales budget and proceeds through a number of steps that ultimately lead to the cash budget, the budgeted income statement, and the budgeted balance sheet.

In recent years, computer-based models and spreadsheet software have been utilized for budgeting in an effort to speed up the budgeting process and allow managerial accountants to investigate the effects of changes in budget assumptions.

Zero-base budgeting (ZBB) has received considerable attention recently as a new approach to budgeting, particularly for use in nonprofit, governmental, and service-type organizations. The chapter discussed the pros and cons of ZBB.

7

COST-VOLUME-PROFIT ANALYSIS AND LEVERAGE

Cost-volume-profit (CVP) analysis, together with cost behavior information, helps engineering managers perform many useful analyses. CVP analysis deals with how profit and costs change with a change in volume. More specifically, it looks at the effects on profits of changes in such factors as variable costs, fixed costs, selling prices, volume, and mix of products sold. By studying the relation ships of costs, sales, and net income, management is better able to cope with many planning decisions.

Break-even analysis, a branch of CVP analysis, determines the break-even sales. The break-even point—the financial crossover point at which revenues exactly match costs—does not show up in corporate earnings reports, but engineers and project managers find it an extremely useful measurement in a variety of ways.

QUESTIONS ANSWERED BY CVP ANALYSIS

CVP analysis tries to answer the following questions:

1. What sales volume is required to break even?
2. What sales volume is necessary to earn a desired profit?
3. What profit can be expected on a given sales volume?
4. How would changes in selling price, variable costs, fixed costs, and output affect profits?
5. How would a change in the mix of products sold affect the break-even, target income volume and profit potential?

CONTRIBUTION MARGIN CONCEPTS

For accurate CVP analysis, a distinction must be made between costs as either variable or fixed. Mixed costs must be separated into their variable and fixed components.

In order to compute the break-even point and perform various CVP analyses, note the following important concepts.

Contribution Margin

The contribution margin (CM) is the excess of sales (S) over the variable costs (VC) of the product or service. It is the amount of money available to cover fixed costs (FC) and to generate profit. Symbolically, $CM = S - VC$.

Unit CM

The unit CM is the excess of the unit selling price (p) over the unit variable cost (v). Symbolically, unit $CM = p - v$.

CM Ratio

The CM ratio is the contribution margin as a percentage of sales, i.e.,

$$CM \text{ ratio} = \frac{CM}{S} = \frac{S - VC}{S} = 1 - \frac{VC}{S}$$

The CM ratio can also be computed using per-unit data as follows:

$$CM \text{ ratio} = \frac{\text{Unit CM}}{p} = \frac{p - v}{p} = 1 - \frac{v}{p}$$

Note that the CM ratio is 1 minus the variable cost ratio. For example, if variable costs are 40% of sales, then the variable cost ratio is 40% and the CM ratio is 60%.

Example 7.1. To illustrate the various concepts of CM, consider the following data for Custom Tool Shop:

	Total	Per unit	Percentage
Sales (1,500 units)	$37,500	$25	100%
Less: Variable costs	15,000	10	40
Contribution margin	$22,500	$15	60%

| Less: Fixed costs | 15,000 |
| Net income | $ 7,500 |

From the data listed above, CM, unit CM, and the CM ratio are computed as:

$$CM = S - VC = \$37,500 - \$15,000 = \$22,500$$

$$\text{Unit CM} = p - v = \$25 - \$10 = \$15$$

$$CM \text{ ratio} = \frac{CM}{S} = \frac{\$22,500}{\$37,500} = 1 - \frac{\$15,000}{\$37,500} = 1 - 0.4 = 0.6 = 60\%$$

or $\quad = \dfrac{\text{Unit CM}}{p} = \dfrac{\$15}{\$25} = 0.6 = 60\%$

ARE YOU BREAKING EVEN?

The break-even point represents the level of sales revenue that equals the total of the variable and fixed costs for a given volume of output at a particular capacity use rate. For example, you might want to ask the break-even occupancy rate (or vacancy rate) for a hotel or the break-even load rate for an airliner.

Generally, the lower the break-even point, the higher the profit and the less the operating risk, other things being equal. The break-even point also provides project managers with insights into profit planning. It can be computed using the following formulas:

$$\text{Break-even point in units} = \frac{\text{Fixed costs}}{\text{Unit CM}}$$

$$\text{Break-even point in dollars} = \frac{\text{Fixed costs}}{\text{CM ratio}}$$

Example 7.2. Using the same data given in Example 7.1, where unit CM = $25 − $10 = $15 and CM ratio = 60%, we get:

Break-even point in units = $15,000/$15 = 1,000 units

Break-even point in dollars = $15,000/0.6 = $25,000

Or, alternatively,

1,000 units × $25 = $25,000

GRAPHICAL APPROACH IN A SPREADSHEET FORMAT

The graphical approach to obtaining the break-even point is based on the so-called *break-even (B-E) chart* which illustrates cost-volume-profit relationships as shown in Figure 7.1. Sales revenue, variable costs, and fixed costs are plotted on the vertical axis while volume, x, is plotted on the horizontal axis. The break-even point is the point where the total sales revenue line intersects the total cost line. The chart can also effectively report profit potentials over a wide range of activity and therefore be used as a tool for discussion and presentation.

The *profit-volume (P-V) chart* as shown in Figure 7.2, focuses directly on how profits vary with changes in volume. Profits are plotted on the vertical axis while units of output are shown on the horizontal axis. The P-V chart provides a quick, condensed comparison of how alternatives of pricing, variable costs, or fixed costs may affect net income as volume changes. The P-V chart can be easily constructed from the B-E chart. Note that the slope of the chart is the unit CM.

DETERMINATION OF TARGET INCOME VOLUME

Besides determining the break-even point, CVP analysis determines the sales required to attain a particular income level or target net income. The formula is:

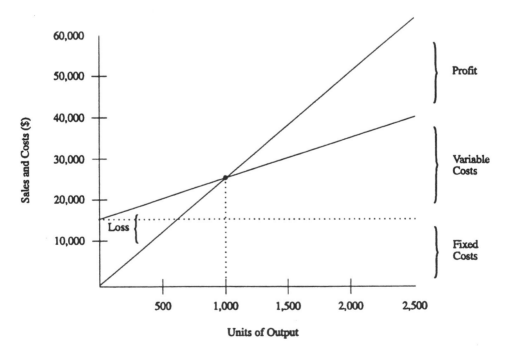

Figure 7.1. Break-Even Chart.

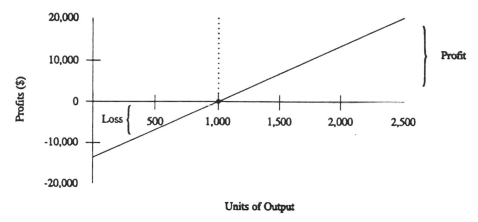

Figure 7.2. Profit-Volume (P-V) Chart.

$$\text{Target income sales volume} = \frac{\text{Fixed costs} + \text{Target income}}{\text{Unit CM}}$$

Example 7.3. Using the same data given in Example 7.1, assume that Custom Tool Shop wishes to attain a target income of $15,000 before tax.

Then, the target income volume would be:

$$\frac{\$15,000 + \$15,000}{\$25 - \$10} = \frac{\$30,000}{\$15} = 2,000 \text{ units}$$

CASH BREAK-EVEN POINT

If a company has a minimum of available cash or the opportunity cost of holding excess cash is too high, management may want to know the volume of sales that will cover all cash expenses during a period. This is known as the cash break-even point. Not all fixed operating costs involve cash payments. For example, depreciation expenses are non-cash fixed charges. To find the cash break-even point, the non-cash charges must be subtracted from the fixed costs. Therefore, the cash break-even point is lower than the usual break-even point. The formula is:

$$\text{Cash break-even point} = \frac{\text{Fixed costs} - \text{Depreciation}}{\text{Unit CM}}$$

Example 7.4. Assume from Example 7.1 that the total fixed costs of $15,000 include depreciation of $1,500. Then the cash break-even point is:

$$\frac{\$15,000 - \$1,500}{\$25 - \$10} = \frac{\$13,500}{\$15} = 900 \text{ units}$$

Custom Tool Shop has to sell 900 units to cover only the fixed costs involving cash payments of $13,500 and to break even.

IMPACT OF INCOME TAXES

If target income is given on an after-tax basis, the target income volume formula becomes:

$$\text{Target income volume} = \frac{\text{Fixed costs} + [\text{target after-tax income}/(1 - \text{tax rate})]}{\text{Unit CM}}$$

Example 7.5. Assume in Example 7.1 that Custom Tool Shop wants to achieve an after-tax income of $6,000. The tax rate is 40%. Then,

$$\text{Target income volume} = \frac{\$15,000 + [\$6,000/(1 - 0.4)]}{\$15}$$

$$= \frac{\$15,000 + \$10,000}{\$15} = 1,667 \text{ units}$$

MARGIN OF SAFETY

The margin of safety is a measure of the difference between the actual sales and the break-even sales. It is the amount by which sales revenue may drop before losses begin, and is expressed as a percentage of expected sales:

$$\text{Margin of safety} = \frac{\text{Expected sales} - \text{Break-even sales}}{\text{Expected sales}}$$

The margin of safety is used as a measure of operating risk. The larger the ratio, the safer the situation since there is less risk of reaching the break-even point.

Example 7.6. Assume Custom Tool Shop projects sales of $35,000 with a break-even sales level of $25,000. The projected margin of safety is

$$\frac{\$35,000 - \$25,000}{\$35,000} = 28.57\%$$

SOME APPLICATIONS OF CVP ANALYSIS AND WHAT-IF ANALYSIS

The concepts of contribution margin and the contribution income statement have many applications in profit planning and short-term decision making. Many "what-if" scenarios can be evaluated using them as planning tools, especially utilizing a spreadsheet program such as *Lotus 1-2-3*. Some applications are illustrated in Examples 7.7 to 7.11 using the same data as in Example 7.1.

Example 7.7. Recall from Example 7.1 that Custom Tool Shop has a CM of 60% and fixed costs of $15,000 per period. Assume that the shop expects sales to go up by $10,000 for the next period. How much will income increase?

Using the CM concepts, we can quickly compute the impact of a change in sales on profits. The formula for computing the impact is:

Change in net income = Dollar change in sales × CM ratio

Thus

Increase in net income = $10,000 × 60% = $6,000

Therefore, the income will go up by $6,000, assuming there is no change in fixed costs.

If we are given a change in unit sales instead of dollars, then the formula becomes:

Change in net income = Change in unit sales × Unit CM

Example 7.8. Assume that the store expects sales to go up by 400 units. How much will income increase? From Example 1, the store's unit CM is $15. Again, assuming there is no change in fixed costs, the income will increase by $6,000.

400 units × $15 = $6,000

Example 7.9. What net income is expected on sales of $47,500?
The answer is the difference between the CM and the fixed costs:

CM: $47,500 × 60%	$28,500
Less: Fixed costs	15,000
Net income	$13,500

Example 7.10. Custom Tool Shop is considering increasing the advertising budget by $5,000, which would increase sales revenue by $8,000. Should the advertising budget be increased?

Income Statement—Contribution Format

	(A) Present (1,500 units)	(B) Proposed (2,400 units)	(B – A) difference
Sales	$37,500 (@$25)	$48,000 (@$20)	$10,500
Less: Variable cost	15,000	24,000	9,000
CM	$22,500	$24,000	$ 1,500
Less: Fixed costs	15,000	16,000	1,000
Net income	$ 7,500	$ 8,000	$ 500

The answer is no, since the increase in the CM is less than the increased cost:

Increase in CM: $8,000 × 60%	$4,800
Increase in advertising	5,000
Decrease in net income	$ (200)

Example 7.11. Consider the original data. Assume again that Custom Tool Shop is currently selling 1,500 units per period. In an effort to increase sales, management is considering cutting its unit price by $5 and increasing the advertising budget by $1,000.

If these two steps are taken, management feels that unit sales will go up by 60%. Should the two steps be taken? The answer is yes. It can be obtained by developing comparative income statements in a contribution format as shown.

SALES MIX ANALYSIS

Break-even and cost-volume-profit analysis requires some additional computations and assumptions when a company produces and sells more than one product. In multi-product firms, sales mix is an important factor in calculating an overall company break-even point.

Different selling prices and different variable costs result in different unit CM and CM ratios. As a result, the break-even points and cost-volume-profit relationships vary with the relative proportions of the products sold, called the *sales mix*.

In break-even and CVP analysis, it is necessary to predetermine the sales mix and then compute a weighted average unit CM. It is also necessary to assume that the sales mix does not change for a specified period. The break-even formula for the company as a whole is:

$$\text{Break-even sales in units (or in dollars)} = \frac{\text{Fixed costs}}{\text{Weighted average unit CM}}$$
$$\text{(or CM ratio)}$$

Example 7.12. Assume that Knibex, Inc. produces cutlery sets made of high-quality wood and steel. The company makes a deluxe cutlery set and a standard set that have the following unit CM data:

	Deluxe	Standard
Selling price	$15	$10
Variable cost per unit	12	5
Unit CM	$ 3	$ 5
Sales mix	60%	40%
Fixed costs		$76,000

The weighted average unit CM = ($3)(0.6) + ($5)(0.4) = $3.80. Therefore, the company's break-even point in units is

$76,000/$3.80 = 20,000 units

which is divided as follows:

A: 20,000 units × 60% = 12,000 units
B: 20,000 units × 40% =   8,000 units
                         20,000 units

Example 7.13. Assume that Dante, Inc. is a producer of recreational equipment. It expects to produce and sell three types of sleeping bags—the Economy, the Regular, and the Backpacker. Information on the bags is given below:

	Economy	Budgeted Regular	Backpacker	Total
Sales	$30,000	$60,000	$10,000	$100,000
Sales mix	30%	60%	10%	100%
Less: VC	24,000	40,000	5,000	69,000
CM	$ 6,000	$20,000	$ 5,000	$ 31,000
CM ratio	20%	33 1/3%	50%	31%
Fixed costs				$ 18,600
Net income				$ 12,400

The CM ratio for Dante, Inc. is $31,000/$100,000 = 31%. Therefore, the break-even point in dollars is

$18,600/0.31 = $60,000

which will be split in the mix ratio of 3:6:1 to give us the following break-even points for the individual products:

Economy: $60,000 × 30% = $18,000

Regular: $60,000 × 60% = 36,000

Backpacker: $60,000 × 10% =   6,000
                            ─────────
                              $60,000

One of the most important assumptions underlying CVP analysis in a multi-product firm is that the sales mix will not change during the planning period. If the sales mix changes, the break-even point will also change.

The shift in sales mix toward the less profitable line, Economy, has caused the CM ratio for the company as a whole to drop from 31% to 26.83%.

Example 7.14. Assume that total sales from Example 7.13 was achieved at $100,000 but that an actual mix came out differently from the budgeted mix (i.e., for Economy, 30% to 55%, for Regular, 60% to 40%, and for Backpacker, 10% to 5%), as shown.

	Economy	Actual Regular	Backpacker	Total
Sales	$55,000	$40,000	$5,000	$100,000
Sales mix	55%	40%	5%	100%
Less: VC	44,000	26,667[a]	2,500[b]	73,167
CM	$11,000	$13,333	$2,500	$ 26,833
CM ratio	20%	33 1/3%	50%	26.83%
Fixed costs				$ 18,600
Net income				$ 8,233

[a]$26,667 = $40,000 × (100% − 33 1/3%) = $40,000 × 66 2/3%
[b]$2,500 = $5,000 × (100% − 50%) = $5,000 × 50%

The new break-even point will be

$18,600/0.2683 = $69,325

The break-even dollar volume has increased from $60,000 to $69,325.

The deterioration (improvement) in the mix caused net income to go down (up). It is important to note that generally, the shift of emphasis from low-margin products to high-margin ones will increase the overall profits of the company.

COST-VOLUME-REVENUE ANALYSIS AND NON-PROFIT ORGANIZATIONS

Cost-volume-profit (CVP) analysis and break-even analysis is not limited to profit firms. CVP is appropriately called *cost-volume-revenue (CVR) analysis*, as it pertains to non-profit organizations. The CVR model not only calculates the break-even service level, but helps answer a variety of "what-if" decision questions.

Example 7.15. LMC, Inc., a Los Angeles county agency, has a $1,200,000 lump-sum annual budget appropriation for an agency to help rehabilitate work-stressed clients. In addition to this, the agency charges each patient $600 a month for board and care. All of the appropriation and revenue must be spent. The variable costs for rehabilitation activity average $700 per patient per month. The agency's annual fixed costs are $800,000. The agency manager wishes to know how many clients can be served. Let x = number of clients to be served.

$$\text{Revenue} = \text{Total expenses}$$

$$\text{Lump sum appropriation} + \$600(12)x = \text{Variable expenses} + \text{Fixed costs}$$

$$\$1,200,000 + \$7,200x = \$8,400x + \$800,000$$

$$(\$7,200 - \$8,400)x = \$800,000 - \$1,200,000$$

$$-\$1,200x = -\$400,000$$

$$x = \$400,000/\$1,200$$

$$x = 333 \text{ clients}$$

We will investigate the following two "what-if" scenarios:

1. Suppose the manager of the agency is concerned that the total budget for the coming year will be cut by 10% to a new amount of $1,080,000. All

other things remain unchanged. The manager wants to know how this budget cut affects next year's service level.

$$\$1,080,000 + \$7,200x = \$8,400x + \$800,000$$
$$(\$7,200 - \$8,400)x = \$800,000 - \$1,080,000$$
$$- \$1,200x = - \$280,000$$
$$x = \$280,000/\$1,200$$
$$x = 233 \text{ clients}$$

2. The manager does not reduce the number of clients served despite a budget cut of 10%. All other things remain unchanged. How much more does he/she have to charge his/her clients for board and care?

In this case, x = board and care charge per year

$$\$1,080,000 + 333x = \$8,400 \ (333) + \$800,000$$
$$333x = \$2,797,200 + \$800,000 - \$1,080,000$$
$$333x = \$2,517,200$$
$$x = \$2,517,200/333 \text{ clients}$$
$$x = \$7,559$$

Thus, the monthly board and care charge must be increased to $630 ($7,559/12 months).

ASSUMPTIONS UNDERLYING BREAK-EVEN AND CVP ANALYSIS

The basic break-even and CVP models are subject to a number of limiting assumptions. They are:

1. The selling price per unit is constant throughout the entire relevant range of activity.
2. All costs are classified as fixed or variable.
3. The variable cost per unit is constant.
4. There is only one product or a constant sales mix.
5. Inventories do not change significantly from period to period.
6. Volume is the only factor affecting variable costs.

LEVERAGE

Leverage is that portion of the fixed costs which represents a risk to the firm. As sales rise, fixed costs do not rise, so profits rise rapidly. On the other hand, as sales decline, these costs remain, so profits fall sharply. There are two major types of leverage—operating and financial.

Operating leverage, a measure of operating risk, refers to the fixed operating costs found in the firm's income statement. Financial leverage, a measure of financial risk, refers to financing a portion of the firm's assets, bearing fixed financing charges in hopes of increasing the return to the common stockholders. The higher the financial leverage, the higher the financial risk. Total leverage is a measure of total risk.

Operating leverage is a measure of operating risk and arises from fixed operating costs. It measures how earnings change due to a change in sales. The formula is:

Operating leverage at a given level of sales (x)

$$= \frac{\text{Percentage change in EBIT}}{\text{Percentage change in sales}}$$

$$= \frac{\Delta \text{EBIT/EBIT}}{\Delta x/x} = \frac{(p - v)\Delta x \, /(p - v)x - FC}{\Delta x/x}$$

$$= \frac{(p - v)x}{(p - v)x - FC}$$

where

EBIT = earnings before interest and taxes (operating income in accounting)

$$= (p - v)x - FC$$

Example 7.16. The Wayne Company manufactures and sells doors to home builders. The doors are sold for $25 each. Variable costs are $15 per door, and fixed operating costs total $50,000. Assume that the Wayne Company is currently selling 6,000 doors per year.

Its operating leverage is:

$$\frac{(p - v)x}{(p - v)x - FC} = \frac{(\$25 - \$15)(6,000)}{(\$25 - \$15)(6,000) - \$50,000} = \frac{\$60,000}{\$10,000} = 6$$

which means if sales increase (decrease) by 1%, the company can expect its net income to increase (decrease) by six times that amount, or 6%.

Financial Leverage

Financial leverage is a measure of financial risk and arises from fixed financial costs. One way to measure financial leverage is to determine how earnings per share (EPS) are affected by a change in EBIT (or operating income).

Financial leverage at a given level of sales (x)

$$= \frac{\text{Percentage in change in EPS}}{\text{Percentage in change in EBIT}} = \frac{(p - v)x - FC}{(p - v)x - FC - IC}$$

where IC is fixed finance charges, i.e., interest expense or preferred stock dividends. [Preferred stock dividend must be adjusted for taxes i.e., preferred stock dividend/$(1 - t)$.]

Example 7.17. Using the data in Example 7.16, the Wayne Company has total financial charges of $2,000, half in interest expense and half in preferred stock dividend. Assume a corporate tax rate of 40%.

First, the fixed financial charges are:

$$IC = \$1,000 + \frac{\$1,000}{(1 - 0.4)} = \$1,000 + \$1,667 = \$2,667$$

Therefore, Wayne's financial leverage is computed as follows:

$$\frac{(p - v)x - FC}{(p - v)x - FC - IC} = \frac{(\$25 - \$15)(6,000) - \$50,000}{(\$25 - \$15)(6,000) - \$50,000 - \$2,667} = \frac{\$10,000}{\$7,333} = 1.36$$

which means that if EBIT increases (decreases) by 1%, Wayne can expect its EPS to increase (decrease) by 1.36 times, or by 1.36%.

Rule of OPM

The rule of OPM (Other People's Money) states that it only pays to borrow if the interest is less than the return on the amount borrowed. In other words, only when the return from invested funds exceeds the cost of financing those funds, the owners of the firm benefit, thus allowing them to profit from the use of other people's money. For example, assume that a firm is able to earn an after-tax return of 12% on its assets. If that firm can borrow at a 10% interest rate in order to expand its assets (assuming a tax rate of 40%), then the shareholders can benefit from favorable leverage. Specifically, the borrowed funds invested in the business will earn an after-tax return of 10%, but the after-tax interest cost of the borrowed funds will be only 6% [10% × (1 − 0.4) = 6%]. The difference will go to the shareholders.

Example 7.18. To further demonstrate the interrelationship between a firm's financial structure and the return it generates on the stockholders' investments, let us compare two firms that generate $300,000 in EBIT (operating income). Both firms employ $800,000 in total assets, but they have different capital structures. One firm employs no debt, whereas the other uses $400,000 in borrowed funds. The comparative capital structures are shown as:

	A	B
Total assets	$800,000	$800,000
Total liabilities	—	400,000
Stockholders' equity (a)	800,000	400,000
Total liabilities and stockholders' equity	$800,000	$800,000

Firm B pays 10% interest for borrowed funds. The comparative income statements and ROEs for firms A and B would look as illustrated.

The absence of debt allows firm A to register higher profits after taxes. Yet the owners in firm B enjoy a significantly higher return on their investments. This provides an important view of the positive contribution debt can make to a business within a certain limit.

But too much debt can increase the firm's financial risk and thus the cost of financing. Unfortunately, leverage is a two-edged sword. If the assets are unable to earn a high enough rate to cover the interest costs of debt, then shareholders suffer.

Comparative Income, ROE

	A	B
Operating income	$300,000	$300,000
Interest expenses	—	(40,000)
Profit before taxes	$300,000	$260,000
Taxes (30% assumed)	(90,000)	(78,000)
Net Profit after taxes (b)	$210,000	$182,000
ROE [(b) / (a)]	26.25%	45.5%

Total Leverage

Total leverage is a measure of total risk. The way to measure total leverage is to determine how EPS is affected by a change in sales.

$$\text{Total leverage at a given level of sales (X)} = \frac{\text{Percentage in change in EPS}}{\text{Percentage in change in sales}}$$

$$= \text{Operating leverage} \times \text{Financial leverage}$$

$$= \frac{(p - v)x}{(p - v)x - FC} \times \frac{(p - v)x - FC}{(p - v)x - FC - IC}$$

$$= \frac{(p - v)x}{(p - v)x - FC - IC}$$

Example 7.19. From Examples 7.16 and 7.17, the total leverage for the Wayne Company is:

Operating leverage x financial leverage = $6 \times 1.36 = 8.16$

or

$$\frac{(p - v)X}{(p - v)x - FC - IC} = \frac{(\$25 - \$15)(6,000)}{(\$25 - \$15)(6,000) - \$50,000 - \$2,667}$$

$$= \frac{\$60,000}{\$7,333} = 8.18 \text{ (due to rounding error)}$$

which means that if sales increase (decrease) by 1%, Wayne can expect its EPS to increase (decrease) by 8.18%.

CONCLUSION

Cost-volume-profit analysis is useful as a frame of reference, as a vehicle for expressing overall managerial performance, and as a planning device via break-even techniques and "what-if" scenarios. Closely related to CVP analysis is the concept of leverage. Leverage is that portion of the fixed costs which represents a risk to the firm.

The following points highlight the analytical usefulness of CVP analysis as a tool for profit planning:

1. A change in either the selling price or the variable cost per unit alters CM or the CM ratio and thus the break-even point.
2. As sales exceed the break-even point, a higher unit CM or CM ratio will result in greater profits than a small unit CM or CM ratio.
3. The lower the break-even sales, the less risky the business and the safer the investment, other things being equal.
4. A large margin of safety means lower operating risk since a large decrease in sales can occur before losses are experienced.
5. Using the contribution income statement model and a spreadsheet program such as Lotus 1-2-3, a variety of "what-if" planning and decision scenarios can be evaluated.
6. In a multi-product firm, sales mix is often more important than overall market share. The emphasis on high-margin products tends to maximize overall profits of the firm.
7. The use of leverage is *favorable* if the return from invested capital exceeds the interest rate on borrowed funds.

We also discussed how the traditional CVP analysis can be applied to the non-profit setting. Illustrations were provided.

8

RESPONSIBILITY ACCOUNTING AND COST CONTROL THROUGH STANDARD COSTS

Responsibility accounting is the system for collecting and reporting revenue and cost information by areas of responsibility. It operates on the premise that managers should be held responsible for their performance, the performance of their subordinates, and all activities within their responsibility center. Responsibility accounting, also called profitability accounting and activity accounting, has the following advantages:

1. It facilitates delegation of decision making.
2. It helps management promote the concept of management by objective. In management by objective, managers agree on a set of goals. The manager's performance is then evaluated based on his or her attainment of these goals.
3. It provides a guide to the evaluation of performance and helps to establish standards of performance which are then used for comparison purposes.
4. It permits effective use of the concept of management by exception, which means that the manager's attention is concentrated on the important deviations from standards and budgets.

RESPONSIBILITY ACCOUNTING AND RESPONSIBILITY CENTER

For an effective responsibility accounting system, the following three basic conditions are necessary:

1. The organization structure must be well defined. Management responsibility and authority must go hand in hand at all levels and must be clearly established and understood.

2. Standards of performance in revenues, costs, and investments must be properly determined and well defined.
3. The responsibility accounting reports (or performance reports) should include only items that are controllable by the manager of the responsibility center. Also, they should highlight items calling for managerial attention.

A well-designed responsibility accounting system establishes responsibility centers within the organization. A responsibility center is defined as a unit in the organization which has control over costs, revenues, and/or investment funds. Responsibility centers can be one of the following types:

Cost Center

A cost center is the unit within the organization which is responsible only for costs. Examples include production and maintenance departments of a manufacturing company. *Variance analysis* based on standard costs and flexible budgets would be a typical performance measure of a cost center.

Profit Center

A profit center is the unit which is held responsible for the revenues earned and costs incurred in that center. Examples might include a sales office of a publishing company, an appliance department in a retail store, and an auto repair center in a department store. The contribution approach to cost allocation is widely used to measure the performance of a profit center.

Investment Center

An investment center is the unit within the organization which is held responsible for the costs, revenues, and related investments made in that center. The corporate headquarters or division in a large decentralized organization would be an example of an investment center.

Figures 8.1 and 8.2 illustrate the manner in which responsibility accounting can be used within an organization and highlights profit and cost centers. This chapter discusses in detail how the performance of both cost and profit centers are evaluated. Performance evaluation of the investment center will be discussed in Chapter 9.

STANDARD COSTS AND VARIANCE ANALYSIS

One of the most important phases of responsibility accounting is establishing standard costs and evaluating performance by comparing actual costs with standard costs. *Standard costs* are costs that are established in advance to serve as

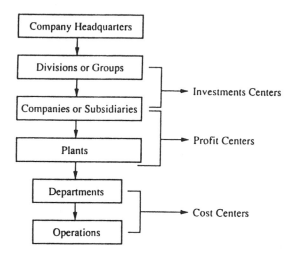

Figure 8.1. Responsibility center within a company.

targets to be met and after the fact, to determine how well those targets were actually met. The standard cost is based on physical and dollar measures. It is determined by multiplying the standard quantity of an input by its standard price.

The difference between the actual costs and the standard costs, called the variance, is calculated for individual cost centers. Variance analysis is a key tool for measuring the performance of a cost center.

The *performance reports* based on the analysis of variances must be prepared for each cost center, addressing the following questions:

1. Is it favorable or unfavorable?
2. If it is unfavorable, is it significant enough for further investigation? (e.g., 5% over the standard)
3. If it is significant, then is it controllable?
4. Who is responsible for what portion of the total variance?
5. What are the causes for an unfavorable variance?
6. What is the remedial action to take?

The report is useful in two ways: (1) it focuses attention on situations in need of management action and (2) it increases the precision of planning and control of costs. The report should be produced as part of the overall standard costing and responsibility accounting system.

GENERAL MODEL FOR VARIANCE ANALYSIS

Two general types of variances can be calculated for most cost items: a price variance and a quantity variance.

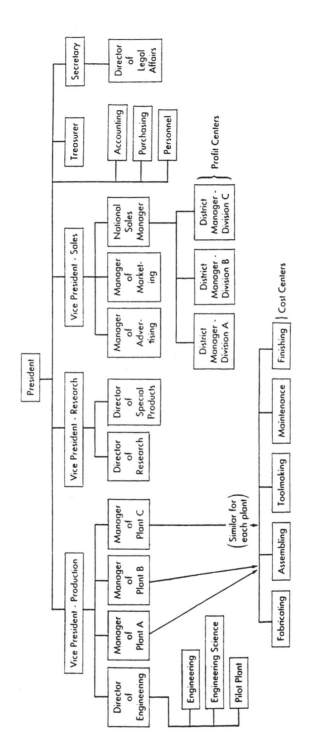

Figure 8.2. Organization chart of a company.

The price variance is calculated as follows:

Price variance = Actual quantity × (Actual price − Standard price)

$$= AQ \times (AP - SP)$$

$$= (AQ \times AP) - (AQ \times SP)$$
$$\quad\quad (1) \quad\quad\quad\quad (2)$$

The quantity variance is calculated as follows:

Quantity   Actual   Standard   Standard
variance = (quantity − quantity) ×   price

$$= (AQ - SQ) \times SP$$

$$= (AQ \times SP) - (SQ \times SP)$$
$$\quad\quad (2) \quad\quad\quad\quad (3)$$

Figure 8.3 shows a general model (3-column model) for variance analysis that incorporates items (1), (2), and (3) from the above equations. It is important to note four things:

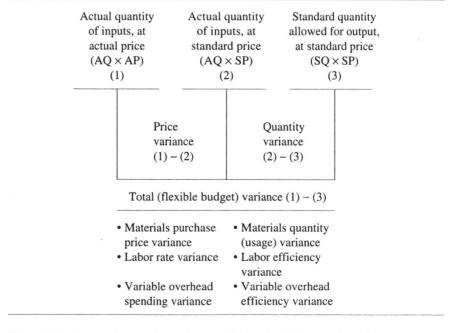

Actual quantity of inputs, at actual price (AQ × AP) (1)	Actual quantity of inputs, at standard price (AQ × SP) (2)	Standard quantity allowed for output, at standard price (SQ × SP) (3)
	Price variance (1) − (2)	Quantity variance (2) − (3)

Total (flexible budget) variance (1) − (3)

• Materials purchase price variance	• Materials quantity (usage) variance
• Labor rate variance	• Labor efficiency variance
• Variable overhead spending variance	• Variable overhead efficiency variance

Figure 8.3. A general model for variance analysis of variable manufacturing costs.

1. A price variance and a quantity variance can be calculated for all three variable cost items—direct materials, direct labor, and the variable portion of factory overhead. The variances are not called by the same name, however. For example, a price variance is called a materials price variance in the case of direct materials, a labor rate variance in the case of direct labor and a variable overhead spending variance in the case of variable factory overhead.
2. A cost variance is unfavorable (U) if the actual price AP or actual quantity AQ exceeds the standard price SP or standard quantity SQ; a variance is favorable (F) if the actual price or actual quantity is less than the standard price or standard quantity.
3. The standard quantity allowed for output—item (3)—is the key concept in variance analysis. This is the standard quantity that should have been used to produce the actual output. It is computed by multiplying the actual output by the number of input units allowed.
4. Variances for fixed overhead are of questionable usefulness for control purposes, since these variances are usually beyond the control of the production department.

We will now illustrate the variance analysis for each of the variable manufacturing cost items.

Materials Variances

A materials purchase price variance is isolated at the time of purchase of the material. It is computed based on the actual quantity purchased. The purchasing department is responsible for any materials price variance that might occur. The materials quantity (usage) variance is computed based on the actual quantity used. The production department is responsible for any materials quantity variance.

Unfavorable price variances may be caused by: inaccurate standard prices, inflationary cost increases, scarcity in raw material supplies resulting in higher prices, and purchasing department inefficiencies. Unfavorable material quantity variances may be explained by poorly trained workers, by improperly adjusted machines, or by outright waste on the production line. Table 8.1 provides the reasons and responsible parties for unfavorable materials variance.

Example 8.1. Houston Corporation uses a standard cost system. The standard variable costs for product J are as follows:

Materials: 2 pounds at $3 per pound
Labor: 1 hour at $5 per hour
Variable overhead: 1 hour at $3 per hour

Table 8.1 Reason and Responsible Party for an Unfavorable Materials Variance

Reason	Responsible party
Overstated price paid, failure to take discounts, improper specifications, insufficient quantities, use of a lower-grade material purchased to economize on price, uneconomical size of purchase orders, failure to obtain an adequate supply of a needed variety, purchase at an irregular time, or sudden and unexpected purchase required	Purchasing
Poor mix of materials, poorly trained workers, improperly adjusted machines, substitution of nonstandard materials, poor production scheduling, poor product design or production technique, lack of proper tools or machines, carelessness in not returning excess materials to storeroom, or unexpected volume changes	Production manager
Failure to detect defective goods	Receiving
Inefficient labor, poor supervision, or waste on the production line	Foreman
Inaccurate standard price	Budgeting
Excessive transportation charges or too small a quantity purchased	Traffic management
Insufficient quantity bought because of a lack of funds	Financial

During March, 25,000 pounds of material were purchased for $74,750 and 20,750 pounds of material were used in producing 10,000 units of finished product. Direct labor costs incurred were $49,896 (10,080 direct labor hours) and variable overhead costs incurred were $34,776.

Using the general model (3-column model), the materials variances are shown in Figure 8.4.

It is important to note that the amount of materials purchased (25,000 pounds) differs from the amount of materials used in production (20,750 pounds). The materials purchase price variance was computed using the 25,000 pounds purchased, while the materials quantity (usage) variance was computed using the 20,750 pounds used in production. A total variance cannot be computed because of the difference.

Alternatively, we can compute the materials variances as follows:

$$\text{Materials purchase price variance} = AQ\,(AP - SP)$$
$$= (AQ \times AP) - (AQ \times SP)$$

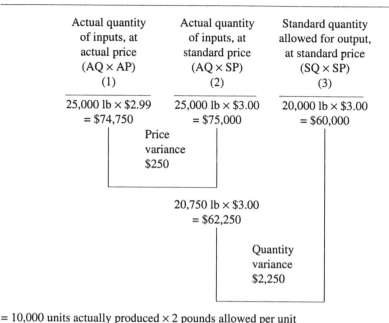

= 10,000 units actually produced × 2 pounds allowed per unit
= 20,000 pounds.

Figure 8.4. Materials variances.

$$= (25,000 \text{ pounds}) (\$2.99 - \$3.00)$$
$$= \$74,750 - \$75,000$$
$$= \$250 \text{ (F)}$$

Materials quantity (usage) variance $= (AQ - SQ)SP$
$$= (20,750 \text{ pounds} - 20,000 \text{ pounds})$$
$$(\$3.00)$$
$$= \$62,250 - \$60,000$$
$$= \$2,250(U)$$

Labor Variances

Labor variances are isolated when labor is used for production. They are computed in a manner similar to the materials variances, except that in the 3-column model the terms efficiency and rate are used in place of the terms quantity and price. The production department is responsible for both the prices paid for labor services and the quantity of labor services used. Therefore, the production department must explain why any labor variances occur.

Unfavorable rate variances may be explained by an increase in wages, or the use of labor commanding higher wage rates than contemplated. Unfavorable efficiency variances may be explained by poor supervision, poor quality workers, poor quality of materials requiring more labor time, machine breakdowns and employee unrest. Table 8.2 provides the causes and responsible parties for unfavorable labor variance.

Example 8.2. Using the same data given in Example 8.1, the labor variances can be calculated as shown in Figure 8.5. Alternatively, we can calculate the labor variances as follows:

$$
\begin{aligned}
\text{Labor rate variance} &= \text{AH (AR} - \text{SR)} \\
&= (\text{AH} \times \text{AR}) - (\text{AH} \times \text{SR}) \\
&= (10,080 \text{ hours})(\$4.95 - \$5.00) \\
&= \$49,896 - \$50,400 \\
&= \$504(\text{F})
\end{aligned}
$$

$$
\begin{aligned}
\text{Labor efficiency variance} &= (\text{AH} - \text{SH})\text{SR} \\
&= (10,080 \text{ hours} - 10,000 \text{ hours}) \times \$5.00
\end{aligned}
$$

Table 8.2 Cause and Responsible Party for an Unfavorable Labor Variance

Reason	Responsible party
Use of overpaid or excessive number of workers	Production manager or union contract
Poor job descriptions or excessive wages	Personnel
Overtime and poor scheduling of production	Production planning
Poor-quality workers or poor training	Personnel or training
Inadequate supervision, inefficient flow of materials, wrong mixture of labor for a given job, inferior tools or idle time from production delays.	Foreman
Employee unrest	Personnel or foreman
Improper functioning of equipment	Maintenance
Insufficient material supply or poor quality	Purchasing

Actual hours of inputs, at actual rate (AH × AR) (1)	Actual hours of inputs, at standard rate (AH × SR) (2)	Standard hours allowed for output, at standard rate (SH × SR) (3)
10,080 h × $4.95 = $49,896	10,080 h × $5.00 = $50,400	10,080 h~ × $5.00 = $50,000

Rate variance (1) – (2) $504(F)	Efficiency variance (2) – (3) $400(U)

Total Variance $104(F)

10,000 units actually produced × 1 hour (h) allowed per unit
= 20,000 pounds.

The symbols AQ, SQ, AP, and SP have been changed to AH, SH, AR, and SR to reflect the terms "hour" and "rate".

Figure 8.5. Labor variances.

$$= \$50,400 - \$50,000$$
$$= \$400(U)$$

Variable Overhead Variances

The variable overhead variances are computed in a way very similar to the labor variances. The production department is usually responsible for any variable overhead variance. Unfavorable variable overhead spending variances may be caused by a large number of factors: acquiring supplies for a price different from the standard, using more supplies than expected, waste, and theft of supplies. Unfavorable variable overhead efficiency variances might be caused by such factors as: poorly trained workers, poor-quality materials, faulty equipment, work interruptions, poor production scheduling, poor supervision, employee unrest, and so on.

When variable overhead is applied using direct labor hours, the efficiency variance will be caused by the same factors that cause the labor efficiency variance. However, when variable overhead is applied using machine hours, inefficiency in machinery will cause a variable overhead efficiency variance.

Actual hours of inputs, at actual rate (AH × AR) (1)	Actual hours of inputs, at standard rate (AH × SR) (2)	Standard hours allowed for output, at standard rate (SH × SR) (3)
10,080 h × $3.45 = $34,776	10,080 h × $3.00 = $30,240	10,080 h × $3.00 = $30,000

Spending variance (1) – (2) $4,536(U)

Efficiency variance (2) – (3) $240(U)

Total Variance $4,776(U)

10,000 units actually produced × 1 hour (h) allowed per unit = 10,000 hours.

Figure 8.6. Variable overhead variances.

Example 8.3. Using the same data given in Example 8.1, the variable overhead variances can be computed as shown in Figure 8.6.

Alternatively, we can compute the variable overhead variances as follows:

$$
\begin{aligned}
\text{Variable overhead spending variance} &= AH\,(AR - SR) \\
&= (AH \times AR) - (AH \times SR) \\
&= (10{,}080 \text{ hours})(\$3.45 - \$3.00) \\
&= \$34{,}776 - \$30{,}240 \\
&= \$4{,}536(U)
\end{aligned}
$$

$$
\begin{aligned}
\text{Variable overhead efficiency variance} &= (AH - SH)SR \\
&= (10{,}080 \text{ hours} - 10{,}000 \text{ hours}) \times \$3.00 \\
&= \$30{,}240 - \$30{,}000 \\
&= \$240(U)
\end{aligned}
$$

FLEXIBLE BUDGETS AND PERFORMANCE REPORTS

A flexible budget is a tool that is extremely useful in cost control. In contrast to a *static (fixed)* budget, which was discussed in Chapter 5, the flexible budget is characterized as follows:

1. It is geared toward a range of activity rather than a single level of activity.
2. It is dynamic in nature rather than static. By using the cost-volume formula (or flexible budget formula), a series of budgets can be easily developed for various levels of activity.

The static budget is geared for only one level of activity and has problems with cost control. Flexible budgeting distinguishes between fixed and variable costs, thus allowing for a budget which can be automatically adjusted (via changes in variable cost totals) to the particular level of activity actually attained. Thus, variances between actual costs and budgeted costs are adjusted for volume ups and downs before differences due to price and quantity factors are computed.

The primary use of the flexible budget is to accurately measure performance by comparing actual costs for a given output with the budgeted costs for the same level of output.

Example 8.5. To illustrate the difference between the static budget and the flexible budget, assume that the Mixing Department of Jones Industries, Inc. is budgeted to produce 6,000 units during June. Assume further that the company was able to produce only 5,800 units. The budget for direct labor and variable overhead costs is as follows:

JONES INDUSTRIES, INC.
The Direct Labor and Variable Overhead Budget
Mixing Department
For the Month of June

Budgeted production	6,000 units
Actual production	5,800 units
Direct labor	$39,000
Variable overhead costs:	
Indirect labor	6,000
Supplies	900
Repairs	300
	$46,200

JONES INDUSTRIES, INC.
The Direct Labor and Variable Overhead Budget
Mixing Department
For the Month of June

	Budget	Actual	Variance (U or F)[a]
Production in units	6,000	5,800	200U
Direct labor	$39,000	$38,500	$500F
Variable overhead costs:			
Indirect labor	6,000	5,950	50F
Supplies	900	870	30F
Repairs	300	295	5F
	$46,200	$45,615	$585F

[a]A variance represents the deviation of actual cost from the standard or budgeted cost. U and F stand for "unfavorable" and "favorable," respectively.

If a static budget approach is used, the performance report will appear as above.

These cost variances are useless, in that they are comparing oranges with apples. The problem is that the budget costs are based on an activity level of 6,000 units, whereas the actual costs were incurred at an activity level below this (5,800 units). From a control standpoint, it makes no sense to try to compare

JONES INDUSTRIES, INC.
Performance Report
Mixing Department
For the Month of June

Budgeted production 6,000 units
Actual production 5,800 units

	Cost-volume formula	Budget 5,800 units	Actual 5,800 units	Variance (U or F)
Direct labor	$6.50 per unit	$37,700	$38,500	$800U
Variable overhead:				
Indirect labor	1.00	5,800	5,950	150U
Supplies	.15	870	870	0
Repairs	.05	290	295	5U
	$7.70	$44,660	$45,615	$955U

costs at one activity level with costs at another activity level. Such comparisons would make a production manager look good as long as the actual production is less than the budgeted production. Using the cost-volume formula and generating the budget based on the 5,800 actual units gives the shown performance report.

Notice that all cost variances are unfavorable (U), as compared to the favorable cost variances on the performance report based on the static budget approach.

Fixed Overhead Variances

By definition, fixed overhead does not change over a relevant range of activity; the amount of fixed overhead per unit varies inversely with the level of production. In order to calculate variances for fixed overhead, it is necessary to determine a standard fixed overhead rate, which requires the selection of a predetermined (denominator) level of activity. This activity should be measured on the basis of standard inputs allowed. The formula is:

$$\text{Standard fixed overhead rate} = \frac{\text{Budgeted fixed overhead}}{\text{Budgeted level of activity}}$$

Total fixed overhead variance is simply under- or over-applied overhead. It is the difference between actual fixed overhead incurred and fixed overhead applied to production (generally, on the basis of standard direct labor hours allowed for actual production). The total fixed overhead variance combines fixed overhead spending (flexible-budget) variance and fixed overhead volume (capacity) variance.

1. *Fixed overhead spending (flexible-budget) variance.* This is the difference between actual fixed overhead incurred and budgeted fixed overhead. This variance is not affected by the level of production. Fixed overhead, by definition, does not change with the level of activity. The spending (flexible-budget) variance is caused solely by events such as unexpected changes in prices and unforeseen repairs.

2. *Fixed overhead volume (capacity) variance.* This variance occurs when the actual level of activity differs from the denominator activity used in determining the standard fixed overhead rate. Note that the denominator used in the formula is the expected annual activity level. Fixed overhead volume variance is a measure of the cost of failure to operate at the denominator (budgeted) activity level, and may be caused by such factors as failure to meet sales targets, idleness due to poor scheduling, and machine breakdowns. The volume variance is calculated as follows:

$$\text{Fixed overhead volume variance} = \text{Budgeted fixed overhead}$$
$$- \text{Fixed overhead applied}$$

or = (Denominator activity – Standard hours allowed)
      × Standard fixed overhead rate

When denominator activity exceeds standard hours allowed, the volume variance is unfavorable (U), because it is an index of less-than-denominator utilization of capacity.

It is important to note that there are no efficiency variances for fixed overhead. Fixed overhead does not change regardless of whether productive resources are used efficiently or not. For example, property taxes, insurance and factory rents are not affected by whether production is being carried on efficiently.

Figure 8.7 illustrates the relationship between the various elements of fixed overhead, and the possible variances.

Example 8.4. The Geige Manufacturing Company has the following standard cost of factory overhead at a normal monthly production (denominator) volume of 1,300 direct labor hours:

Variable overhead (1 hour @ $2)
Fixed overhead (1 hour @ $5)

Fixed overhead budgeted is $6,500 per month. During the month of March, the following events occurred:

	Incurred: Actual hours × Actual rate (1)	Flexible budget Based on actual hours (2)	Flexible budget based on standard hours allowed (3)	Applied (4)
3-way analysis	Spending variance (1) – (2)	Efficiency variance (not applicable)	Volume variance (3) – (4)	
2-way analysis	Flexible budget variance (1) – (3)		Volume variance (3) – (4)	
	Under- or over applied (1) – (4)			

Figure 8.7. Fixed overhead variances.

1. Actual overhead costs incurred (for 1,350 hours) were:

    Variable $2,853
    Fixed $6,725

2. Standard hours allowed, 1,250 hours (1 hour × 1,250 units of output)

Note that:

1. Flexible budget formula:

    Variable overhead rate $2 per direct labor hour
    Fixed overhead budgeted $6,500

2. Standard overhead applied rates:

    Variable $2 per direct labor hour
    Fixed $5 per direct labor hour

Figure 8.8 shows all the variances for variable overhead as well as fixed overhead.

Alternatively, fixed overhead volume variance can be calculated as follows:

Fixed overhead volume variance = (Denominator activity − standard hours
                                  allowed) × Standard fixed overhead rate
                               = (1,300 hours − 1,250 hours) × $5
                               = 50 hours × $5 = $250(U)

Methods of Variance Analysis for Factory Overhead

Variance analysis for factory overhead consists of a two-, three-, or four-way method of computation, depending on the significance of the variance amounts compared to the cost of analysis. These methods are indicated in Figure 8.7 and Figure 8.8.

The two-way analysis computes two variances: budget variance (sometimes called the flexible-budget or controllable variance) and volume variance, which means:

1. Budget variance = Variable spending variance + Fixed spending
                              (budget) variance + Variable efficiency variance
2. Volume variance = Fixed volume variance

Incurred: Actual hours × Actual rate (1,350 hrs) (1)	Flexible budget based on actual hours (1,350 hrs) (2)	Flexible budget based on standard hours allowed (1,250 hrs) (3)	Applied (1,250 hrs) (4)
V $2,853	$2,700 (1,350 × $2)	$2,500 (1,250 × $2)	$2,500
F 6,725	6,500	6,500	6,250
$9,578	$9,200	$9,000	$8,750

(3-way)	Spending variance (1) − (2)	Efficiency variance (not applicable)	Volume variance (3) − (4)
	V $153(U)	$200(U)	Not applicable
	F 225(U)	Not applicable	$250(U)
	$378(U)	$200(U)	$250(U)

(2-way)	Flexible budget variance (1) − (3) V $353(U) F 225(U) $578(U)		Volume variance (3) − (4) Not applicable $250(U) $250(U)

Under- or over applied
(1) − (4)
V $353(U)
F 475(U)
$828(U)

Figure 8.8. Variance analysis for variable overhead and fixed overhead.

The three-way analysis computes three variances: spending, efficiency, and volume variances. Therefore,

1. Spending variance = Variable spending variance + Fixed spending (budget) variance
2. Efficiency variance = Variable efficiency variance
3. Volume variance = Fixed volume variance

The four-way analysis includes the following:

1. Variable spending variance
2. Fixed spending (budget) variance
3. Variable efficiency variance
4. Fixed volume variance

CONCLUSION

Variance analysis is essential in the organization for the appraisal of all aspects of the business. This chapter was concerned with the control of cost centers through standard costs. It discussed the basic mechanics of how the two major variances—the price variance and the quantity variance—are calculated for direct materials, direct labor, variable overhead, and fixed overhead. Also presented were the managerial significance of these variances. The idea of flexible budgeting was emphasized in an attempt to correctly measure the efficiency of the cost center. We noted that fixed overhead volume variance has limited usefulness at the level of a cost center, since only top management has the power to expand or contract fixed facilities.

9

IMPROVING DIVISIONAL PERFORMANCE

The ability to measure performance is essential in developing management incentives and controlling the operation toward the achievement of organizational goals. A typical decentralized subunit is an investment center which is responsible for an organization's invested capital (operating assets) and the related operating income.

There are two widely used measurements of performance for the investment center: the rate of return on investment (ROI) and residual income (RI).

Goods and services are often exchanged between various divisions of a decentralized organization. The transfer price is the selling price credited to the selling division and the cost charged to the buying division for an internal transfer of a good or service. The choice of transfer prices not only affects divisional performance but is also important in decisions involving make or buy, whether to buy internally or outside, and choosing between production possibilities.

RATE OF RETURN ON INVESTMENT (ROI)

ROI relates net income to invest capital. Specifically,

$$ROI = \frac{\text{Operating income}}{\text{Operating assets}}$$

Example 9.1. Consider the following financial data for a division:

Operating assets $100,000
Operating income $18,000
ROI = $18,000/$100,000 = 18%

The problem with this formula is that it only indicates how a division did and how well it fared in the company. Other than that, it has very little value from the standpoint of profit planning.

THE BREAKDOWN OF ROI—DUPONT FORMULA

In the past, managers have tended to focus only on the margin earned and have ignored the turnover of assets. It is important to realize that excessive funds tied up in assets can be just as much of a drag on profitability as excessive expenses.

Dupont was the first major company to recognize the importance of looking at both margin and asset turnover in assessing the performance of an investment center. The ROI breakdown, known as the *Dupont formula*, is expressed as a product of these two factors, as shown below.

$$\text{ROI} = \frac{\text{Operating income}}{\text{Operating assets}} = \frac{\text{Operating income}}{\text{Sales}} \times \frac{\text{Sales}}{\text{Operating assets}}$$

$$= \text{Margin} \times \text{Asset turnover}$$

The Dupont formula combines the income statement and balance sheet into this otherwise static measure of performance. Margin is a measure of profitability or operating efficiency. It is the percentage of profit earned on sales. This percentage shows how many cents are attached to each dollar of sales. On the other hand, asset turnover measures how well a division manages its assets. It is the number of times by which the investment in assets turn over each year to generate sales.

The breakdown of ROI is based on the theory that the profitability of a firm is directly related to management's ability to manage assets efficiently and to control expenses effectively.

Example 9.2. Assume the same data as in Example 9.1. Also assume sales of $200,000.

$$\text{Then, ROI} = \frac{\text{Operating income}}{\text{Operating assets}} = \frac{\$18,000}{\$200,000} = 18\%$$

Alternatively,

$$\text{Margin} = \frac{\text{Operating income}}{\text{Sales}} = \frac{\$18,000}{\$200,000} = 9\%$$

$$\text{Turnover} = \frac{\text{Sales}}{\text{Operating assets}} = \frac{\$200,000}{\$100,000} = 2 \text{ times}$$

Therefore,

ROI = Margin × Turnover = 9% × 2 times

The breakdown provides a lot of insight to division managers on how to improve profitability of the investment center. Specifically, it has several advantages over the original formula for profit planning. They are:

1. Focusing on the breakdown of ROI provides the basis for integrating many of the management concerns that influence a division's overall performance. This will help managers gain an advantage in the competitive environment.
2. The importance of turnover as a key to overall return on investment is emphasized in the breakdown. In fact, turnover is just as important as profit margin in enhancing overall return.
3. The importance of sales is explicitly recognized, which it is not in the original formula.
4. The breakdown stresses the possibility of trading between margin and turnover in an attempt to improve the overall performance of a company. The margin and turnover complement each other. In other words, a low turnover can be made up for by a high margin; and vice versa.

Example 9.3. The breakdown of ROI into its two components shows that a number of combinations of margin and turnover can yield the same rate of return, as shown below:

Margin	×	Turnover	= ROI
(1) 9%	×	2 times	= 18%
(2) 6	×	3	= 18
(3) 3	×	6	= 18
(4) 2	×	9	= 18

The turnover-margin relationship and its resulting ROI is depicted in Figure 9.1.

ROI And Profit Planning

The breakdown of ROI into margin and turnover gives divisional managers insight into planning for profit improvement by revealing where weaknesses exist: in margin or turnover, or both. Various actions can be taken to enhance ROI. Generally, they can:

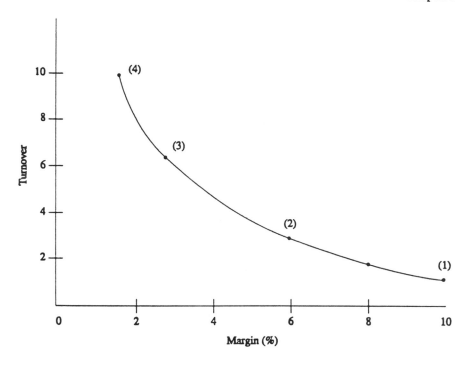

Figure 9.1 The margin-turnover relationship.

1. Improve margin
2. Improve turnover
3. Improve both

Alternative 1 demonstrates a popular way of improving performance. Margins may be increased by reducing expenses, raising selling prices, or increasing sales faster than expenses. Some of the ways to reduce expenses are:

1. Use less costly inputs of materials.
2. Automate processes as much as possible to increase labor productivity.
3. Bring the discretionary fixed costs under scrutiny, with various programs either curtailed or eliminated. Discretionary fixed costs arise from annual budgeting decisions by management. Examples include advertising, research and development, and management development programs. The cost-benefit analysis is called for in order to justify the budgeted amount of each discretionary program.

A division with pricing power can raise selling prices and retain profitability without losing business. Pricing power is the ability to raise prices even in poor economic times when unit sales volume may be flat and capacity may not be fully utilized. It is also the ability to pass on cost increases to consumers without attracting domestic and import competition, political opposition, regulation, new entrants, or threats of product substitution. The division with pricing power must have a unique economic position. Divisions that offer unique, high-quality goods and services (where the service is more important than the cost) have this economic position.

Alternative 2 may be achieved by increasing sales while holding the investment in assets relatively constant, or by reducing assets. Some of the strategies to reduce assets are:

1. Dispose of obsolete and redundant inventory. The computer has been extremely helpful in this regard, making perpetual inventory methods more feasible for inventory control.
2. Devise various methods of speeding up the collection of receivables and also evaluate credit terms and policies.
3. See if there are unused fixed assets.
4. Use the converted assets obtained from the use of the previous methods to repay outstanding debts or repurchase outstanding issues of stock. The division may release them elsewhere to get more profit, which will improve margin as well as turnover.

Alternative 3 may be achieved by increasing sales or by any combinations of alternatives 1 and 2.

Figure 9.2 shows complete details of the relationship of ROI to the underlying ratios—margin and turnover—and their components. This will help identify more detailed strategies to improve margin, turnover, or both.

Example 9.4. Assume that management sets a 20% ROI as a profit target. It is currently making an 18% return on its investment.

$$\text{ROI} = \frac{\text{Operating income}}{\text{Operating assets}} = \frac{\text{Operating income}}{\text{Sales}} \times \frac{\text{Sales}}{\text{Operating assets}}$$

Present situation:

$$18\% = \frac{18{,}000}{200{,}000} \times \frac{200{,}000}{100{,}000}$$

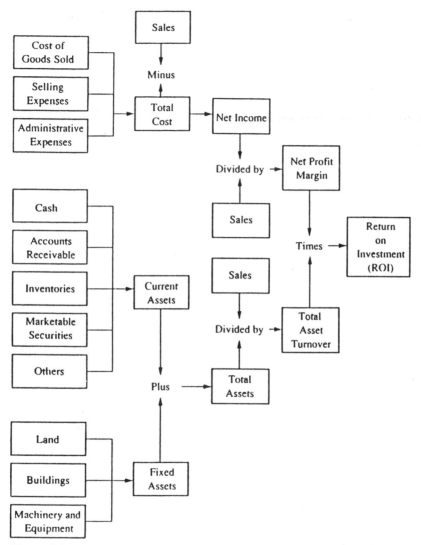

Figure 9.2 Relationships of factors influencing ROI.

The following are illustrative of the strategies which might be used (each strategy is independent of the other).

Alternative 1: Increase the margin while holding turnover constant. Pursuing this strategy would involve leaving selling prices as they are and making every effort to increase efficiency so as to reduce expenses. By doing so, expenses might be reduced by $2,000 without affecting sales and investment to yield a 20% target ROI, as follows:

$$20\% = \frac{20,000}{200,000} \times \frac{200,000}{100,000}$$

Alternative 2: Increase turnover by reducing investment in assets while holding net profit and sales constant. Working capital might be reduced or some land might be sold, reducing investment in assets by $10,000 without affecting sales and net income to yield the 20% target ROI as follows:

$$20\% = \frac{18,000}{200,000} \times \frac{200,000}{90,000}$$

Alternative 3: Increase both margin and turnover by disposing of obsolete and redundant inventories or through an active advertising campaign. For example, trimming down $5,000 worth of investment in inventories would also reduce the inventory holding charge by $1,000. This strategy would increase ROI to 20%.

$$20\% = \frac{19,000}{200,000} \times \frac{200,000}{95,000}$$

Excessive investment in assets is just as much of a drag on profitability as excessive expenses. In this case, cutting unnecessary inventories also helps cut down expenses of carrying those inventories, so that both margin and turnover are improved at the same time. In practice, alternative 3 is much more common than alternatives 1 or 2.

RESIDUAL INCOME (RI)

Another approach to measuring performance in an investment center is residual income (RI). RI is the operating income which an investment center is able to earn above some minimum rate of return on its operating assets. RI, unlike ROI, is an absolute amount of income rather than a specific rate of return. When RI is used to evaluate divisional performance, the objective is to maximize the total amount of residual income, not to maximize the overall ROI figure.

RI = Operating income – (Minimum required rate of return × Operating assets)

Example 9.5. In Example 9.1, assume the minimum required rate of return is 13%. Then, the residual income of the division is

$$\$18,000 - (13\% \times \$100,000) = \$18,000 - \$13,000 = \$5,000$$

RI is regarded as a better measure of performance than ROI because it encourages investment in projects that would be rejected under ROI. A major disadvantage of RI, however, is that it cannot be used to compare divisions of different sizes. RI tends to favor the larger divisions due to the larger amount of dollars involved.

INVESTMENT DECISIONS UNDER ROI AND RI

The decision whether to use ROI or RI as a measure of divisional performance affects financial managers' investment decisions. Under the ROI method, division managers tend to accept only the investments whose returns exceed the division's ROI; otherwise, the division's overall ROI would decrease. Under the RI method on the other hand, division managers would accept an investment as long as it earns a rate in excess of the minimum required rate of return. The addition of such an investment will increase the division's overall RI.

Example 9.6. Consider the same data given in Examples 9.1 and 9.2:

Operating assets	$100,000
Operating income	$18,000
Minimum required rate of return	13%
ROI = 18% and RI = $5,000	

Assume that the division is presented with a project that would yield 15% on a $10,000 investment. The division manager would not accept this project under the ROI approach since the division is already earning 18%. Acquiring this project will bring down the present ROI to 17.73%, as shown below:

	Present	New project	Overall
Operating assets (a)	$100,000	$10,000	$110,000
Operating income (b)	18,000	1,500[*]	19,500
ROI (b / a)	18%	15%	17.73%

[*]$10,000 × 15% = $1,500

Under the RI approach, the manager would accept the new project since it provides a higher rate than the minimum required rate of return (15% vs. 13%).

Accepting the new project will increase the overall residual income to $5,200, as shown below:

	Present	New project	Overall
Operating assets (a)	$100,000	$10,000	$110,000
Operating income (b)	18,000	1,500	19,500
Minimum required income at 13% (c)	13,000	1,300*	14,300
RI (b – a)	$ 5,000	$ 200	$ 5,200

*$10,000 × 13% = $1,300

TRANSFER PRICING

Goods and services are often exchanged between various divisions of a decentralized organization. A major goal of transfer pricing is to enable divisions that exchange goods or services to act as independent businesses.

The question then is: What monetary values should be assigned to these exchanges or transfers? Market price? Some kind of cost? Some version of either? Unfortunately, there is no single transfer price that will please everybody —that is, top management, the selling division, and the buying division involved in the transfer. Various transfer pricing schemes are available, such as market price, cost-based price, or negotiated price.

The choice of a transfer pricing policy (i.e., which type of transfer price to use) is normally decided by top management. The decision will typically include consideration of the following:

Goal congruence. Will the transfer price promote the goals of the company as a whole? Will it harmonize the divisional goals with organizational goals?

Performance evaluation. Will the selling division receive enough credit for its transfer of goods and services to the buying division? Will the transfer price hurt the performance of the selling division?

Autonomy. Will the transfer price preserve autonomy—the freedom of the selling and buying division managers to operate their divisions as decentralized entities?

Other factors such as minimization of tariffs and income taxes and observance of legal restrictions.

Transfer prices can be based on:

Market price
Cost-based price—variable or full cost

Negotiated price
General formula, which is usually the sum of variable costs per unit and
   opportunity cost for the company as a whole (lost revenue per unit on
   outside sales)

Market Price

Market price is the best transfer price in the sense that it will maximize the profits of the company as a whole, if it meets the following two conditions:

1. There exists a competitive market price.
2. Divisions are independent of each other.

If either one of these conditions is violated, market price will not lead to an
optimal economic decision for the company.

Cost-Based Price—Variable or Full Cost

Cost-based transfer price, another alternative transfer pricing scheme, is easy to
understand and convenient to use. But there are some disadvantages, including:

Inefficiencies of selling divisions are passed on to the buying divisions with
   little incentive to control costs. The use of standard costs is recom-
   mended in such a case.
The cost-based method treats the divisions as cost centers rather than profit
   or investment centers. Therefore, measures such as ROI and RI cannot
   be used for evaluation purposes.

The variable-cost-based transfer price has an advantage over the full cost
method because in the short run it may tend to ensure the best utilization of the
overall company's resources. The reason is that, in the short run, fixed costs do
not change. Any use of facilities, without incurrence of additional fixed costs,
will increase the company's overall profits.

Negotiated Price

A negotiated price is generally used when there is no clear outside market. A
negotiated price is a price agreed upon between the buying and selling divisions
that reflects unusual or mitigating circumstances. This method is widely used
when no intermediate market price exists for the product transferred and the
selling division is assured of a normal profit.

Example 9.7. Company X just purchased a small company that specializes in the manufacture of part no. 323. Company X is a decentralized organization, and will treat the newly acquired company as an autonomous division called Division B with full profit responsibility.

Division B's fixed costs total $30,000 per month, and variable costs per unit are $18. Division B's operating capacity is 5,000 units. The selling price per unit is $30. Division A of Company X is currently purchasing 2,500 units of part no. 323 per month from an outside supplier at $29 per unit, which represents the normal $30 price less a quantity discount.

Top management of the company wishes to decide what transfer price should be used.

Top management may consider the following alternative prices:

1. $30 market price
2. $29, the price that Division A is currently paying to the outside supplier
3. $23.50 negotiated price, which is $18 variable cost plus 1/2 of the benefits of an internal transfer [($29 − $18) × 1/2]
4. $24 full cost, which is $18 variable cost plus $6 ($30,000/5,000 units) fixed cost per unit
5. $18 variable cost

We will discuss each of these prices:

1. $30 would not be an appropriate transfer price. Division B cannot charge a price more than the price Division A is paying now ($29).
2. $29 would be an appropriate transfer price if top management wishes to treat the divisions as autonomous investment centers. This price would cause all of the benefits of internal transfers to accrue to the selling division, with the buying division's position remaining unchanged.
3. $23.50 would be an appropriate transfer price if top management wishes to treat the divisions as investment centers, but wishes to share the benefits of an internal transfer equally between them, as follows.

Variable costs of Division B	$18.00
1/2 of the difference between the variable costs of Division B and the price Division A is paying ($29 − $18) × 1/2	5.50
Transfer price	$23.50

Note that $23.50 is just one example of a negotiated transfer price. The exact price depends on how the benefits are divided.

4. $24 [$24 = $18 + ($30,000 / 5,000 units)] would be an appropriate transfer price if top management treats the divisions like cost centers with no

profit responsibility. All benefits from both divisions will accrue to the buying division. This will maximize the profits of the company as a whole, but adversely affect the performance of the selling division. Another disadvantage of this cost-based approach is that inefficiencies (if any) of the selling division are being passed on to the buying division.

5. $18 would be an appropriate transfer price for guiding top management in deciding whether transfers between the two divisions should take place. Since $18 is less than the outside purchase price of the buying division, and the selling division has excess capacity, the transfer should take place, because it will maximize the profits of the company as a whole. However, if $18 is used as a transfer price, then all of the benefits of the internal transfer accrue to the buying division and it will hurt the performance of the selling division.

General Formula

It is not easy to find a cure-all answer to the transfer pricing problem, since the three problems of goal congruence, performance evaluation, and autonomy must all be considered simultaneously. It is generally agreed, however, that some form of competitive market price is the best approach to the transfer pricing problem. The following formula would be helpful in this effort:

Transfer price =   Variable costs per unit + Opportunity costs per unit for the company as a whole

Opportunity costs are defined here as net revenue given up by the company as a whole if the goods and services are transferred internally. The reasoning behind this formula is that the selling division should be allowed to recover its variable costs plus opportunity cost (i.e., revenue that it could have made by selling to an outsider) of the transfer. The selling department should not have to suffer lost income by selling within the company.

Example 9.8. Company X has more than 50 divisions, including A, B, and K. Division A, the buying division, wants to buy a component for its final product and has an option to buy from Division B or from an outside supplier at the market price of $200. If Division A buys from the outside supplier, it will in turn buy selected raw materials from Division K for $40. This will increase its contribution to overall company profits by $30 ($40 revenue minus $10 variable costs). Division B, on the other hand, can sell its component to Division A or to an outside buyer at the same price. Division B, working at full capacity, incurs variable costs of $150. Will the use of $200 as a transfer price lead to optimal decisions for the company as a whole? Figure 9.3 depicts the situation.

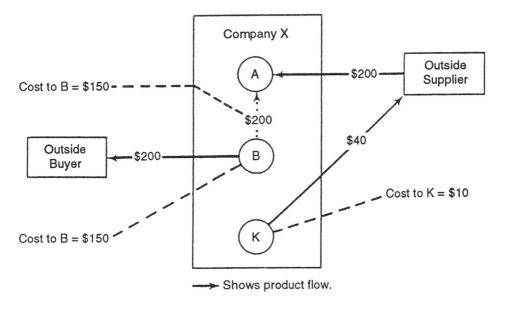

Figure 9.3 Transfer pricing situation.

The optimal decision from the viewpoint of Company X as a whole can be looked at in terms of its net cash outflow, as follows:

	Division A's action	
	Buy from B	Buy from outsider
Outflow to the company as a whole	$(150)	$(200)
Cash inflows	—	to B:$50($200–$150)
		to K:$30($40–$10)
Net cash outflow to the company as a whole	$(150)	$(120)

To maximize the profits of Company X, Division A should buy from an outside supplier. The transfer price that would force division A to buy outside should be the sum of variable costs and opportunity costs, that is,

$150 + $50 + $30 = $230 per unit

In other words, if Division B charges $230 to Division A, division A will definitely buy from the outside source for $200.

CONCLUSION

Return on investment (ROI) and residual income (RI) are the two most widely used measures of divisional performance. Emphasis was placed on the breakdown of the ROI formula, commonly referred to as the *Du Pont formula*. The breakdown formula has several advantages over the original formula in terms of profit planning. The choice of evaluation systems—ROI or RI—will greatly affect a division's investment decisions.

Transfer pricing relates to the price to be charged in an exchange of goods and services between two investment centers within an organization. Unfortunately, there is no single transfer price that is satisfactory to the parties involved in the transfer—the selling division, buying division, and top management. In practical terms, the best transfer price to use is negotiated market price. However, when that is not available, for example when there is a new product, budgeted cost plus profit markup should be used. In any event, the buying division should never be charged a transfer price that exceeds the outside market price. Whether the buying division is allowed to buy outside or stay inside depends on what is best for corporate profitability.

10

RELEVANT COSTING IN NONROUTINE DECISIONS

When performing the manufacturing and selling functions, engineering management is constantly faced with the problem of choosing between alternative courses of action. Typical questions to be answered include: What to make? How to make it? Where to sell the product? and, what price should be charged? In the short run, management is faced with many short-term, nonroutine decisions. In a short-term situation, fixed costs are generally irrelevant to the decision at hand. Engineering managers must recognize as major decision tools the two important concepts: *relevant costs* and *contribution margin.*

RELEVANT COSTS DEFINED

In each of the above situations, the ultimate management decision rests on cost data analysis. Cost data are important in many decisions, since they are the basis for profit calculations. Cost data are classified by function, behavior patterns, and other criteria, as previously discussed.

However, not all costs are of equal importance in decision making, and managers must identify the costs that are relevant to a decision. Such costs are called relevant costs. The relevant costs are the expected future costs (and also revenues) which differ between the decision alternatives. Therefore, the sunk costs (past and historical costs) are not considered relevant in the decision at hand. What is relevant are the incremental or differential costs.

Under the concept of relevant costs, which may be appropriately titled the incremental, differential, or relevant cost approach, the decision involves the following steps:

1. Gather all costs associated with each alternative.
2. Drop the sunk costs.

3. Drop those costs which do not differ between alternatives.
4. Select the best alternative based on the remaining cost data.

TYPES OF DECISIONS

Pricing a Special Order

A company often receives a short-term, special order for its products at lower prices than usual. During normal times, the company may refuse such an order since it will not yield a satisfactory profit. If times are bad however, such an order should be accepted if the incremental revenue obtained from it exceeds the incremental costs. The company is better off receiving some revenue, above its incremental costs, than to receive nothing at all.

Such a price, one lower than the regular price, is called a *contribution price*. This approach to pricing is often called the contribution approach to pricing or the variable pricing model.

This approach is most appropriate under the following conditions:

1. when operating in a distress situation
2. when there is idle capacity
3. when faced with sharp competition or in a competitive bidding situation

Example 10.1. Assume that a company with 100,000 unit capacity is currently producing and selling only 90,000 units of product each year at a regular price of $2. If the variable cost per unit is $1 and the annual fixed cost is $45,000, the income statement looks as follows:

Sales (90,000 units)	$180,000	$2.00
Less: Variable cost		
(90,000 units)	90,000	1.00
Contribution margin	$ 90,000	$1.00
Less: Fixed cost	45,000	0.50
Net income	$ 45,000	$0.50

The company has just received an order that calls for 10,000 units @ $1.20, for a total of $12,000. The acceptance of this order will not affect regular sales. The company's president is reluctant to accept the order, however, because the $1.20 price is below the $1.50 factory unit cost ($1.50 = $1.00 + $0.50). Should the company accept the order?

The answer is yes. The company can add to total profits by accepting this special order even though the price offered is below the unit factory cost. At a price of $1.20, the order will contribute $0.20 per unit (CM per unit = $1.20 −

	Per unit	Without special order (90,000 units)	With special order (100,000 units)	Difference
Sales	$2.00	$180,000	$192,000	$12,000
Less: Variable costs	1.00	90,000	100,000	10,000
CM	$1.00	$ 90,000	$ 92,000	$ 2,000
Less: Fixed cost	0.50	45,000	45,000	—
Net income	$0.50	$ 45,000	$ 47,000	$ 2,000

$1.00 = $0.20) toward fixed cost, and profit will increase by $2,000 (10,000 units × $0.20).

Using the contribution approach to pricing, the variable cost of $1 will be a better guide than the full unit cost of $1.50. Note that the fixed costs do not change because of the presence of idle capacity.

The same result can be seen as shown in the special order comparison table.

Example 10.2. The marketing manager had decided that for Product A he wants a markup of 30% over cost. Particulars concerning a unit of Product A are given as follows:

Direct material	$ 4,000
Direct labor	10,000
Overhead	2,500
Total cost	16,500
Markup on cost (30%)	4,950
Selling price	$21,450

Total direct labor for the year equals $1,200,000. Total overhead for the year equals 25% of direct labor ($300,000), of which 40% is fixed and 60% is variable. The customer offers to buy a unit of Product A for $18,000. Idle capacity exists.

The extra order should be accepted because it provides an increased contribution margin, as indicated in the contribution analysis table.

Bid Price

The relevant cost approach can be used to determine the bid price on a contract.

Selling price		$18,000
Less: Variable costs		
Direct material	$4,000	
Direct labor	10,000	
Variable overhead		
($10,000 × 15%)*	1,500	(15,500)
Contribution margin		$ 2,500
Less: fixed overhead		(0)
Net income		$ 2,500

*Variable overhead equals 15% of direct labor, calculated as follows:

$$\frac{\text{Variable overhead}}{\text{Direct labor}} = \frac{60\% \times \$300,000}{\$1,200,000} = \frac{\$180,000}{\$1,200,000} = 15\%$$

Example 10.3. Travis Company has received an order for 6,000 units. The management accountant wants to know the minimum bid price that would produce a $14,000 increase in profit. The current income statement is shown.

If the contract is taken, the cost patterns for the extra order will remain the same, with these exceptions:

Transportation costs will be paid by the customer.
Special tools costing $6,000 will be required for just this order and will not be reusable.
Direct labor time for each unit under the order will be 10% longer.

Income Statement

Sales (30,000 units × $20)		$600,000
Less cost of sales		
Direct material	$60,000	
Direct labor	150,000	
Variable overhead (150,000 × 40%)	60,000	
Fixed overhead	80,000	(350,000)
Gross margin		$250,000
Less selling and administrative expenses		
Variable (includes transportation costs of $0.20 per unit)	15,000	
Fixed	85,000	(100,000)
Net income		$150,000

	Calculating Bid Price	
	Current cost	Per unit
Selling price	$20	($600,000/30,000)
Direct material	2	($60,000/30,000)
Direct labor	5	($150,000/30,000)
Variable overhead		40% of direct labor cost
		($60,000/$150,000)
Variable selling and		
administrative expense	$0.50	($15,000/30,000)

The bid price is derived as shown in the table with the bid price calculation.

As can be seen in the following income statement, the contract price for the 6,000 units should be $80,000 ($680,000 − $600,000), or $13.33 per unit ($80,000/6,000).

The contract price per unit of $13.33 is less than the $20 current selling price per unit. Note, that by accepting the order, total fixed cost will remain the same except for the $6,000 cost of special tools.

Pricing Standard Products

Unlike pricing special orders, pricing standard products requires long-term considerations. Here, the key concept is to recognize that the established unit selling price must be sufficient in the long run to cover all manufacturing, selling, and administrative costs, both fixed and variable, as well as to provide for an adequate return and for future expansion.

There are two primary approaches to pricing standard products that are sold on the regular market: the full cost approach and the contribution approach. Both approaches use some kind of *cost-plus pricing formula.*

1. The full (absorption) cost approach defines the cost base as the full unit manufacturing cost. Selling and administrative costs are provided for through the markup that is added to the cost base.
2. The contribution approach defines the cost base as the unit variable cost. Fixed costs are provided for through the markup that is added to this base.

Example 10.4. The XYZ company has accumulated the following cost data on its regular product:

Income Statement

	Current 30,000 units	Projected 36,000 units
Sales	$600,000	$680,000[d] Computed last
Cost of sales		
Direct material	$ 60,000	$ 72,000 ($2 × 36,000)
Direct labor	150,000	183,000 ($150,000 + [6,000 × $5.50[a]])
Variable overhead	$ 60,000	$ 73,200 ($183,000 × 40%)
Fixed overhead	80,000	86,000 ($80,000 + $6,000)
Total	$350,000	$414,200
Variable selling and administration costs	$ 15,000	$ 16,800 ($15,000 + [6,000 × $0.30])[b]
Fixed selling and administrative costs	85,000	85,000
Total	$100,000	$101,800
Net income	$150,000	$164,000[c]

[a]$5 × 1.10 = $5.50
[b]$0.50 − $0.20 = $0.30
[c]$150,000 + $14,000 = $164,000
[d]Net income + Selling and administrative expenses + Cost of sales = sales
$164,000 + $101,800 + $414,200 = $680,000

	Per unit	Total
Direct materials	$6	
Direct labor	4	
Variable overhead	4	
Fixed overhead (based on 20,000 units)	6	$120,000
Variable selling and admin. expenses	1	
Fixed selling and admin. expenses (based on 20,000 units)	2	40,000

Assume that in order to obtain its desired selling price, the firm has a general policy of adding a markup equal to 50% of the full unit cost or 100% of the unit variable cost.

Under the full cost approach, the desired unit selling price is:

Direct materials	$ 6
Direct labor	4

Factory overhead	10 ($4 + $6)
Full unit cost	$20
Markup to cover selling and admin. expenses, and desired profit—	
50% of full unit cost	10
Desired selling price	$30

Under the contribution approach, the desired selling price is determined as follows:

Direct materials	$ 6
Direct labor	4
Variable costs (overhead, selling and admin.)	5
Unit variable cost	$15
Markup to cover fixed costs and desired profit—100% of unit variable cost	15
Desired selling price	$30

In determining the percentage markup, companies use their desired rate of return as the base.

Example 10.5. Assume that the XYZ Company has determined that a $500,000 investment is necessary to manufacture and market 20,000 units of its product every year. It will cost $20 to manufacture each unit at a 20,000 unit level of activity, and total selling and administrative expenses are estimated to be $100,000. If the company desires a 20% return on investment, what will be the required markup using the full cost approach?

Desired rate of return	
(20% × $500,000)	$100,000
Selling and admin. expenses	100,000
Total	$200,000 (a)
Full unit cost	
(20,000 units × $20)	$400,000 (b)
Required Markup — (a) / (b)	50%

The Make-or-Buy Decision

The decision whether to produce a component part internally or to buy it externally from an outside supplier is called a "make-or-buy" decision. This decision involves both quantitative and qualitative factors. The qualitative factors include ensuring product quality and the necessity for long-term business relationships with the supplier. The quantitative factors deal with cost. The quantitative effects of the make-or-buy decision are best seen through the relevant cost approach.

Example 10.6. Assume that a firm has prepared the following cost estimates for the manufacture of a subassembly component based on an annual production of 8,000 units as shown.

	Per unit	Total
Direct materials	$ 5	$ 40,000
Direct labor	4	32,000
Variable factory overhead applied	4	32,000
Fixed factory overhead applied		
(150% of direct labor cost)	6	48,000
Total cost	$19	$152,000

The supplier has offered to provide the subassembly at a price of $16 each. Two-thirds of fixed factory overhead, which represents executive salaries, rent, depreciation, and taxes, remain constant regardless of the decision. Should the company buy or make the product?

The key to the decision lies in the investigation of those relevant costs that change between the make-or-buy alternatives. Assuming that the productive capacity will be idle if not used to produce the subassembly, the analysis takes the form shown in the make vs. buy table.

	Make vs. Buy			
	Per unit		Total of 8,000 units	
	Make	Buy	Make	Buy
Purchase price		$16		$128,000
Direct materials	$ 5		$ 40,000	
Direct labor	4		32,000	
Variable overhead	4		32,000	
Fixed overhead that can be				
avoided by not making	2		16,000	
Total relevant costs	$15	$16	$120,000	$128,000
Difference in favor of making	$1		$8,000	

The make-or-buy decision must be investigated, along with the broader perspective of considering how best to utilize available facilities. The alternatives are:

1. Leaving facilities idle
2. Buying the parts and renting out idle facilities
3. Buying the parts and using idle facilities for other products

The Sell-or-Process-Further Decision

When two or more products are produced simultaneously from the same input by a joint process, these products are called joint products. The term joint costs is used to describe all the manufacturing costs incurred prior to the point where the joint products are identified as individual products, referred to as the split-off point. At the split-off point some of the joint products are in final form and saleable to the consumer, whereas others require additional processing.

In many cases, however, the company might have an option: it can sell the goods at the split-off point or process them further in the hope of obtaining additional revenue. In connection with this type of decision, called the "sell-or-process-further" decision, joint costs are considered irrelevant, since the joint costs have already been incurred at the time of the decision, and therefore represent sunk costs. The decision will rely exclusively on additional revenue compared to the additional costs incurred due to further processing.

Example 10.7. The Gin Company produces three products, A, B, and C from a joint process. Joint production costs for the year were $120,000. Product A may be sold at the split-off point or processed further. The additional processing requires no special facilities and all additional processing costs are variable. Sales values and costs needed to evaluate the company's production policy regarding product A follow:

Units produced	Sales value at split-off	Additional cost and sales value after further processing	
		Sales	Costs
3,000	$60,000	$90,000	$25,000

Should product A be sold at the split-off point or processed further?

Incremental sales revenue	$30,000
Incremental costs, additional processing	25,000
Incremental gain	$ 5,000

In summary, product A should be processed as shown above. Keep in mind that the joint production cost of $120,000 is not included in the analysis, since it is a sunk cost and therefore, irrelevant to the decision.

Keeping or Dropping A Product Line

The decision whether to drop an old product line or add a new one must take into account both qualitative and quantitative factors. However, any final decision should be based primarily on the impact the decision will have on contribution margin or net income.

Example 10.8. The ABC grocery store has three major product lines: produce, meats, and canned food. The store is considering the decision to drop the meat line because the income statement shows it is being sold at a loss. Note the income statement for these product lines below.

In this example, direct fixed costs are those costs that are identified directly with each of the product lines, whereas allocated fixed costs are the amount of common fixed costs allocated to the product lines using some base such as space occupied. The amount of common fixed costs typically continues regardless of the decision and thus cannot be saved by dropping the product line to which it is distributed.

The comparative approach showing the effects on the company as a whole with and without the meat line is shown in the keep vs. drop meats comparative analysis.

Income Statement—Product Lines

	Produce	Meats	Canned Food	Total
Sales	$10,000	$15,000	$25,000	$50,000
Less: Variable costs	6,000	8,000	12,000	26,000
CM	$ 4,000	$ 7,000	$13,000	$24,000
Less: Fixed costs				
Direct	$ 2,000	$ 6,500	$ 4,000	$12,500
Allocated	1,000	1,500	2,500	5,000
Total $	3,000	$ 8,000	$ 6,500	$17,500
Net income	$ 1,000	$(1,000)	$ 6,500	$ 6,500

	Keep Meats	Drop Meats	Difference
Sales	$50,000	$35,000	$(15,000)
Less: Variable cost	26,000	18,000	(8,000)
CM	$24,000	$17,000	$ (7,000)
Less: Fixed cost			
Direct	$12,500	$ 6,000	$ (6,500)
Allocated	5,000	5,000	—
Total	$17,500	$11,000	$ (6,500)
Net Income	$ 6,500	$ 6,000	$ (500)

Alternatively, the incremental approach would show the following:

If Meats Dropped		
Sales revenue lost		$15,000
Gains:		
Variable cost avoided	$8,000	
Direct Fixed costs avoided	6,500	14,500
Increase (decrease) in net income		$ (500)

From either of the two methods, we see that by dropping meats the store will lose an additional $500. Therefore, the meat product line should be kept. One of the great dangers in allocating common fixed costs is that such allocations can make a product line look less profitable than it really is. Because of such an allocation, the meat line showed a loss of $1,000, but it in effect contributes $500 ($7,000 − $6,500) to the recovery of the store's common fixed costs.

UTILIZATION OF SCARCE RESOURCES

In general, the emphasis on products with higher contribution margins maximizes a firm's total net income, even though total sales may decrease. This is not true, however, when there are constraining factors and scarce resources. The constraining factor may be machine hours, labor hours, or cubic feet of warehouse space. In the presence of these constraining factors, maximizing total profits depends on getting the highest contribution margin per unit of the factor (rather than the highest contribution margin per unit of product output).

Example 10.9. Assume that a company produces two products, A and B, with the following contribution margins per unit.

	A	B
Sales	$8	$24
Variable costs	6	20
CM	$2	$ 4
Annual fixed costs		$42,000

As is indicated by CM per unit, B is more profitable than A since it contributes more to the company's total profits than A ($4 vs. $2). But let us assume that the firm has a limited capacity of 10,000 labor hours. In addition, assume that A requires two labor hours to produce and B requires five labor hours. One way to express this limited capacity is to determine the contribution margin per labor hour.

	A	B
CM/unit	$2.00	$4.00
Labor hours required per unit	2	5
CM per labor hour	$1.00	$0.80

Since A returns the higher CM per labor hour, it should be produced and B should be dropped. Another way to look at the problem is to calculate total CM for each product.

	A	B
Maximum possible production	5,000 units[a]	2,000 units[b]
CM per unit	$ 2	$ 4
Total CM	$10,000	$8,000

[a](10,000 hours/2 hours)
[b](10,000 hours/5 hours)

Again, product A should be produced since it contributes more than B ($10,000 vs. $8,000).

CONCLUSION

Not all costs are of equal importance in decision making, and engineering managers must identify the costs that are relevant to a decision. The relevant costs are the expected future costs that differ between the decision alternatives. Therefore, the sunk costs are irrelevant since they are past and historical costs. The costs that continue regardless of the decision are irrelevant.

What is relevant are the incremental or differential costs. The relevant cost approach assists managerial accountants in making short-term, nonroutine decisions such as whether to accept a below normal selling price, which products to emphasize, whether to make or buy, whether to sell or process further, how to formulate a bid price on a contract, and how to optimize utilization of capacity. Table 10.1 summarizes guidelines for typical short-tern decisions.

Table 10.1 Decision Guidelines

Decision	Description	Decision guidelines
Special order	Should a discount-priced order be accepted when there is idle capacity?	If regular orders are not affected, accept order when the revenue from the order exceeds the incremental cost. Fixed costs are usually irrelevant.
Make or buy	Should a part be made or bought from a vendor?	Choose lower-cost option. Fixed costs are usually irrelevant. Often opportunity costs are present.
Closing a segment	Should a segment be dropped?	Compare loss in contribution margin with savings in fixed costs.
Sell or, process, further	Should joint products be sold at split-off or processed further?	Ignore joint costs. Process further if incremental revenue exceeds incremental cost.
Scarce, resources	Which products should be emphasized when capacity is limited?	Emphasize products with highest contribution margin per unit of scarce resource (e.g., CM per machine hour).

11

APPLYING THE TIME VALUE OF MONEY

A dollar now is worth more than a dollar to be received later. This statement sums up an important principle: money has a time value. The basis of this principle is not that inflation might make the dollar received at a later time worth less in buying power. The reason is that you could invest the dollar now and have more than a dollar at the specified later date.

Time value of money is a critical consideration in financial and investment decisions. For example, compound interest calculations are needed to determine future sums of money resulting from an investment. Discounting, or the calculation of present value, which is inversely related to compounding, is used to evaluate the future cash flow associated with capital budgeting projects. There are plenty of applications of the time value of money in finance.

HOW DO YOU CALCULATE FUTURE VALUES—HOW MONEY GROWS

A dollar in hand today is worth more than a dollar to be received tomorrow because of the interest it could earn from putting it in a savings account or placing it in an investment account. Compounding interest means that interest earns interest. For the discussion of the concepts of compounding and time value, let us define:

F_n = future value: the amount of money at the end of year n
P = Principal
i = Annual interest rate
n = Number of years

Then,

F_1 = The amount of money at the end of year 1

= Principal and Interest = $P + iP = P(1+i)$

F_2 = The amount of money at the end of year 2

= $F_1(1+i) = P(1+i)(1+i) = P(1+i)^2$

The future value of an investment compounded annually at rate i for n years is

$$F_n = P(1+i)^n = P \times T1(i,n)$$

where $T1(i,n)$ is the compound amount of $1 and can be found in Table A1 (Appendix).

Example 11.1. You place $1,000 in a savings account earning 8% interest compounded annually. How much money will you have in the account at the end of 4 years?

$$F_n = P(1+i)^n$$
$$F_4 = \$1,000 \, (1 + 0.08)^4 = \$1,000 \, T1(8\%,4 \text{ years})$$

From Table 1, the T1 for 4 years at 8% is 1.361. Therefore, $F_4 = \$1,000(1.361) = \$1,361$.

Example 11.2. You invested a large sum of money in the stock of TLC Corporation. The company paid a $3 dividend per share. The dividend is expected to increase by 20% per year for the next 3 years. You wish to project the dividends for years 1 through 3.

$$F_n = P(1+i)^n$$
$$F_1 = \$3(1+0.2)^1 = \$3 \, T1(20\%,1) = \$3(1.200) = \$3.60$$
$$F_2 = \$3(1+0.2)^2 = \$3 \, T1(20\%,2) = \$3(1.440) = \$4.32$$
$$F_3 = \$3(1+0.2)^3 = \$3 \, T1(20\%,3) = \$3(1.728) = \$5.18$$

Intrayear Compounding

Interest is often compounded more frequently than once a year. Banks, for example, compound interest quarterly, daily and even continuously. If interest is compounded m times a year, then the general formula for solving the future value becomes

$$F_n = P\left(1 + \frac{i}{m}\right)^{n \times m} = P \times T1(i/m, n \times m)$$

The formula reflects more frequent compounding (n.m) at a smaller interest rate per period (i/m). For example, in the case of semiannual compounding (m = 2), the above formula becomes

$$F_n = P\left(1 + \frac{i}{2}\right)^{n \times 2} = P \times T1(i/2, n \times 2)$$

For continuous compounding of interest e^{in} is the compound factor, where e = 2.71828...,

$$F_n = P\, e^{in}$$

Example 11.3. You deposit $10,000 in an account offering an annual interest rate of 20 percent. You will keep the money on deposit for five years. The interest rate is compounded quarterly. The accumulated amount at the end of the fifth year is calculated as follows:

$$F_n = P\left(1 + \frac{i}{m}\right)^{nm} = P \times T1(i/m, nm)$$

where   P = $10,000
        i/m = 20%/4 = 5%
        nm = 5 × 4 = 20

Therefore,

$$F_5 = \$10,000(1 + .05)^{20} = \$10,000\ T1(5\%, 20)$$
$$= \$10,000\ (2.653) = \$26,530$$

Example 11.4. Assume that P = $1,000. i= 8% and n = 2 years. Then for Annual compounding (m = 1):

$$F_2 = \$1,000(1+0.08)^2 = \$1,000\ T1(8\%, 2)$$
$$= \$1,000(1.166)$$
$$= \$1,166.00$$

Semiannual compounding (m = 2):

$$F_2 = \$1,000\left(1+\frac{0.08}{2}\right)^{2\times2}$$

$$= \$1,000(1+.04)4 = \$1,000 \ T1(4\%,4)$$

$$= \$1,000(1.170)$$

$$= \$1,170.00$$

Quarterly compounding (m = 4):

$$F_2 = \$1,000\left(1+\frac{0.08}{4}\right)^{2\times4}$$

$$= \$1,000(1+.02)8 = \$1,000 \ T1(2\%,8)$$

$$= \$1,000(1.172)$$

$$= \$1,172.00$$

For continuous compounding:

$$F_2 = P \ e^{in} = \$1,000 \ (2.71828)^{0.08 \times 2}$$

$$= \$1,000 \ (2.71828)^{0.16} = \$1,000 \ (1.1735)$$

$$= \$1,173.50$$

As the example shows, the more frequently interest is compounded, the greater the amount accumulated. This is true for any interest for any period of time.

Future Value of an Annuity

An annuity is defined as a series of payments (or receipts) of a fixed amount for a specified number of periods. Each payment is assumed to occur at the end of the period. The future value of an annuity is a compound annuity which involves depositing or investing an equal sum of money at the end of each year for a certain number of years and allowing it to grow.

Let S_n = the future value on an n-year annuity
A = the amount of an annuity

Then we can say

$$S_n = A(1+i)^{n-1} + A(1+i)^{n-2} + \ldots + A(1+i)^0$$

$$= A[(1+i)^{n-1} + (1+i)^{n-2} + \ldots + (1+i)^0]$$

$$= A \times \sum_{t=0}^{n-1} (1 + i)^t = A \frac{(1 + i)^n - 1}{i} = A \times T2(i,n)$$

where $T2(i,n)$ represents the future value of an annuity of $1 for n years compounded at i% and can be found in Table A2, Appendix.

Example 11.5. You wish to determine the sum of money you will have in a savings account at the end of 6 years by depositing $1,000 at the end of each year for the next 6 years. The annual interest rate is 8%. The T2(8%,6 years) is given in Table A2 (Appendix) as 7.336. Therefore,

$S_6 = \$1,000 \ T2(8\%,6) = \$1,000(7.336) = \$7,336$

Example 11.6. You deposit $30,000 semiannually into a fund for ten years. The annual interest rate is 8%. The amount accumulated at the end of the tenth year is calculated as follows:

$S_n = A \times T2(i,n)$

where $A = \$30,000$
$\quad\quad\quad i = 8\%/2 = 4\%$
$\quad\quad\quad n = 10 \times 2 = 20$

Therefore,

$S_n = \$30,000 \ T2(4\%, 20)$
$\quad\quad = \$30,000 \ (29.778) = \$893,340$

Note: The formula for computing the future value of an *annuity due* must take into consideration one additional year of compounding, since the payment comes at the beginning of the year. Therefore, the future value formula must be modified to take this into consideration by compounding it for one more year. The formula is: $A \times T2(i,n)(1 + i)$

Can a Computer Help?

Financial calculators marketed by several manufacturers (e.g., Hewlett-Packard, Sharp, Texas Instruments) have a "future (compound) value" function. Future value is also incorporated as a built-in function in spreadsheet programs such as Lotus 1-2-3. For example, Lotus 1-2-3 has a routine @FV(*payments,interest, term*) which calculates the future value of an investment, based on a series of equal payments. To calculate the future value of an annuity due, use the formula $@FV(payments, interest, term) \times (1 + interest)$.

WHAT IS PRESENT VALUE—HOW MUCH IS THE MONEY WORTH NOW?

Present value is the present worth of future sums of money. The process of calculating present values, or discounting, is actually the opposite of finding the compounded future value. In connection with present value calculations, the interest rate i is called the *discount rate*. The discount rate we use is more commonly called the *cost of capital*, which is the minimum rate of return required by the investor. Determining the cost of capital is covered in detail in Chapter 10 (Cost of Capital).

Recall that $F_n = P(1+i)^n$

Therefore,

$$P = \frac{F_n}{(1+i)^n} = F_n \left[\frac{1}{(1+i)^n} \right] = F_n \times T3(i,n)$$

where T3(i,n) is the present value of $1, given in Table A3, Appendix.

Example 11.7. You have been given an opportunity to receive $20,000 six years from now. If you can earn 10% on your investments, what is the most you should pay for this opportunity? To answer this question, you must compute the present value of $20,000 to be received 6 years from now at a 10% rate of discount. F_6 is $20,000, i is 10%, and n is 6 years. T3(10%,6) from Table A3 is 0.565.

$$P = \$20,000 \left[\frac{1}{(1+0.1)^6} \right] = \$20,000 \, T3(10\%,6) = \$20,000(0.564) = \$11,280$$

This means that you can earn 10% on your investment, and you would be indifferent to receiving $11,280 now or $20,000 6 years from today since the amounts are time equivalent. In other words, you could invest $11,300 today at 10% and have $20,000 in 6 years.

Present Value of Mixed Streams of Cash Flows

The present value of a series of mixed payments (or receipts) is the sum of the present value of each individual payment. We know that the present value of each individual payment is the payment times the appropriate T3 value.

Example 11.8. You are thinking of starting a new product line that initially costs $32,000. Your annual projected cash inflows are:

    1    $10,000
    2    $20,000
    3     $5,000

If you must earn a minimum of 10% on your investment, should you initiate this new product line?

The present value of this series of mixed streams of cash inflows is calculated as follows:

Year	Cash inflows	$\times$ T3(10%,n)	Present value
1	$10,000	0.909	$9,090
2	$20,000	0.826	$16,520
3	$5,000	0.751	$3,755
			$29,365

Since the present value of your projected cash inflows is less than the initial investment, you should not undertake this project.

Present Value of an Annuity

Interest received from bonds, pension funds, and insurance obligations all involve annuities. To compare these financial instruments, we need to know the present value of each. The present value of an annuity (P_n) can be found by using the following equation:

$$P_n = A \times \left[\frac{1}{(1+i)^1}\right] + A \times \left[\frac{1}{(1+i)^2}\right] + \ldots + A \times \left[\frac{1}{(1+i)^n}\right]$$

$$= A\left[\frac{1}{(1+i)^1} + \frac{1}{(1+i)^2} + \ldots + \frac{1}{(1+i)^n}\right]$$

$$= A\sum_{t=1}^{n}\frac{1}{(1+i)^t} = \frac{1}{i}\left[1 - \frac{1}{(1+i)}\right] = A \times T4(i,n)$$

where T4(i,n) represents the present value of an annuity of $1 discounted at i% for n years and is found in Table A4, Appendix.

Example 11.9. Assume that the cash inflows in Example 11.8 form an annuity of $10,000 for 3 years. Then the present value is

$P_n = A \times T4(i,n)$

$P_3 = \$10,000 \ T4(10\%, 3 \text{ years}) = \$10,000 \ (2.487) = \$24,870$

Note: The formula for computing the present value of an *annuity due* must take into consideration one additional year of compounding, since the payment comes at the beginning of the year. Therefore, the present value formula must be modified to take this into consideration by compounding it for one more year. The formula is: $A \times T4(i,n)(1 + i)$

Can A Computer Help?

Again, computer software can be extremely helpful in making present value calculations. For example, @PV(*payments,interest,term*) of Lotus 1-2-3 determines the present value of an investment, based on a series of equal payments, discounted at a periodic interest rate over the number of periods. To calculate the present value of an annuity due, use the following formula: @PV(*payments, interest,term*) $\times$ (1 + *interest*). Financial calculators can do this too.

Perpetuities

Some annuities go on forever and are called perpetuities. An example of a perpetuity is preferred stock which yields a constant dollar dividend indefinitely. The present value of a perpetuity is found as follows:

$$\text{Present value of a perpetuity} = \frac{\text{Receipt}}{\text{Discount rate}} = \frac{A}{i}$$

Example 11.10. Assume that a perpetual bond has an \$80-per-year interest payment and that the discount rate is 10%. The present value of this perpetuity is:

$$P = \frac{A}{i} = \frac{\$80}{0.10} = \$800$$

WHAT ARE THE APPLICATIONS OF FUTURE VALUES AND PRESENT VALUES?

Future and present values have numerous applications in financial and investment decisions. Six of these applications are presented below.

Deposits to Accumulate a Future Sum (or Sinking Fund)

A financial manager might wish to find the annual deposit (or payment) that is necessary to accumulate a future sum. To find this future amount (or sinking fund) we can use the formula for finding the future value of an annuity.

$$S_n = A \times T2(i,n)$$

Solving for A, we obtain:

$$\text{Annual deposit amount} = A = \frac{S_n}{T2(i,n)}$$

Example 11.11. You wish to determine the equal annual end-of-year deposits required to accumulate \$5,000 at the end of 5 years in a fund. The interest rate is 10%. The annual deposit is:

$S_5 = \$5,000$
$T2(10\%, 5 \text{ years}) = 6.105$ (from Table 2, Appendix)

$$A = \frac{\$5,000}{6.105} = \$819$$

In other words, if you deposit \$819 at the end of each year for 5 years at 10% interest, you will have accumulated \$5,000 at the end of the fifth year.

Example 11.12. You need a sinking fund for the retirement of a bond 30 years from now. The interest rate is 10%. The annual year-end contribution needed to accumulate \$1,000,000 is

$S_{30} = \$1,000,000$
$T2(10\%, 30 \text{ years}) = 164.49$

$$A = \frac{\$1,000,000}{164.49} = \$6,079.40$$

Amortized Loans

If a loan is to be repaid in equal periodic amounts, it is said to be an amortized loan. Examples include auto loans, mortgage loans, and most commercial loans. The periodic payment can easily be computed as follows:

$P_n = A \times T4(i,n)$

Solving for A, we obtain:

$$\text{Amount of loan} = A = \frac{P_n}{T4(i,n)}$$

Example 11.13. You borrow $200,000 for five years at an interest rate of 14%. The annual year-end payment on the loan is calculated as follows:

$P_5 = \$200,000$
$T4(14\%, 5 \text{ years}) = 3.433$ (from Table 4)

$$\text{Amount of loan} = A = \frac{P_5}{T4(14\%, 5 \text{ years})} = \frac{\$200,000}{3.433} = \$58,258.08$$

Example 11.14. You take out a 40-month bank loan of $5,000 at a 12% annual interest rate. You want to find out the monthly loan payment.

$i = 12\%/12 \text{ months} = 1\%$
$P_{40} = \$5,000$
$T4(1\%, 40 \text{ months}) = 32.835$ (from Table 4)

$$\text{Therefore, } A = \frac{\$5,000}{32.835} = \$152.28$$

So to repay the principal and interest on a $5,000, 12%, 40-month loan, you have to pay $152.28 a month for the next 40 months.

Example 11.15. Assume that a firm borrows $2,000 to be repaid in three equal installments at the end of each of the next 3 years. The bank charges 12% interest. The amount of each payment is

$P_3 = \$2,000$
$T4(12\%, 3 \text{ years}) = 2.402$

$$\text{Therefore, } A = \frac{\$2,000}{2.402} = \$832.64$$

How to Develop a Loan Amortization Schedule

Each loan payment consists partly of interest and partly of principal. The breakdown is often displayed in a loan amortization schedule. The interest component of the payment is largest in the first period (because the principal balance is the highest) and subsequently declines, whereas the principal portion is smallest in the first period (because of the high interest) and increases thereafter, as shown in the following example.

Example 11.16. Using the same data as in Example 11.15, we set up the following amortization schedule:

Year	Payment	Interest	Repayment of principal	Remaining balance
0	—	—	—	$2,000.00
1	$832.64	$240.00[a]	$592.64[b]	$1,407.36
2	$832.64	$168.88	$663.76	$743.60
3	$832.64	$89.23	$743.41[c]	

a   Interest is computed by multiplying the loan balance at the beginning of the year by the interest rate. Therefore, interest in year 1 is $2,000(0.12) = $240; in year 2 interest is $1,407.36(0.12) = $168.88; and in year 3 interest is $743.60(0.12)= $89.23. All figures are rounded.

b   The reduction in principal equals the payment less the interest portion ($832.64 − $240.00 = $592.64)

c   Not exact because of accumulated rounding errors.

Annual Percentage Rate (APR)

Different types of investments use different compounding periods. For example, most bonds pay interest semiannually; banks generally pay interest quarterly. If a financial manager wishes to compare investments with different compounding periods, he or she needs to put them on a common basis. The annual percentage rate (APR), or effective annual rate, is used for this purpose and is computed as follows:

$$APR = \left(1 + \frac{i}{m}\right)^m - 1.0$$

where i = the stated, nominal or quoted rate and m = the number of compounding periods per year.

Example 11.17. If the nominal rate is 6%, compounded quarterly, the APR is

$$APR = \left(1+\frac{i}{m}\right)^m - 1.0 = \left(1+\frac{0.06}{4}\right)^4 - 1.0 = (1.015)^4 - 1.0 = 1.0614 - 1.0 = 0.0614 = 6.14\%$$

This means that if one bank offered 6% with quarterly compounding, while another offered 6.14% with annual compounding, they would both be paying the same effective rate of interest.

Annual percentage rate (APR) also is a measure of the cost of credit, expressed as a yearly rate. It includes interest as well as other financial charges such as loan origination and certain closing fees. The lender is required to tell you the APR. It provides you with a good basis for comparing the cost of loans, including mortgage plans.

Rates of Growth

In finance, it is necessary to calculate the compound annual rate of growth, associated with a stream of earnings. The compound annual growth rate in earnings per share is computed as follows:

$$F_n = P \times T1(i,n)$$

Solving this for T1, we obtain

$$T1(i,n) = \frac{F_n}{P}$$

Example 11.18. Assume that your company has earnings per share of $2.50 in 19x1, and 10 years later the earnings per share has increased to $3.70. The compound annual rate of growth in earnings per share is computed as follows:

$$F_{10} = \$3.70 \text{ and } P = \$2.50$$

Therefore,

$$T1(i,10) = \frac{\$3.70}{\$2.50} = 1.48$$

From Table 1, Appendix, a T1 of 1.48 at 10 years is at i = 4%. The compound annual rate of growth is therefore 4%.

Bond Values

Bonds call for the payment of a specific amount of interest for a stated number of years and the repayment of the face value at the maturity date. Thus, a bond represents an annuity plus a lump sum. Its value is found as the present value of the payment stream. The interest is usually paid semiannually.

$$V = \sum_{t=1}^{n} \frac{I}{(1 + i)^t} + \frac{M}{(1 + i)^n}$$
$$= I \times T4(i,n) + M \times T3(i,n)$$

where

I = Interest payment per period
M = Par value, or maturity value, usually $1,000
i = Investor's required rate of return
n = Number of periods

Example 11.19. Assume there is a 10-year bond with a 10% coupon, paying interest semiannually and having a face value of $1,000. Since interest is paid semiannually, the number of periods involved is 20 and the semiannual cash inflow is $100/2 = $50.

Assume that you have a required rate of return of 12% for this type of bond. Then, the present value (V) of this bond is:

$$V = \$50 \times T4(6\%, 20) + \$1,000 \times T3(6\%, 20)$$
$$= \$50(11.470) + \$1,000(0.312) = \$573.50 + \$312.00 = \$885.50$$

Note that the required rate of return (12%) is higher than the coupon rate of interest (10%), and so the bond value (or the price investors are willing to pay for this particular bond) is less than its $1,000 face value.

Use of Financial Calculators and Spreadsheet Programs

There are many financial calculators that contain pre-programmed formulas to perform many present value and future applications. Furthermore, spreadsheet software such as Lotus 1-2-3, Microsoft's Excel, and Borland's Quattro Pro has built-in financial functions to perform many of these applications. For example, @PMT(*principal,interest,term*) in Lotus 1-2-3 calculates the amount of the periodic payment to pay off a loan, given a specified periodic interest rate and number of payment periods.

CONCLUSION

The basic idea of the time value of money is that money received in the future is not as valuable as money received today. The time value of money is a critical factor in many financial and investment applications such as finding the amount of deposits needed to accumulate a future sum and the periodic payment of an amortized loan. The development of the time value of money concept permits comparison of sums of money that are available at different points in time. This chapter developed two basic concepts: future value and present value. It showed how these values are calculated and can be applied to various financial and investment situations.

12

EVALUATING CAPITAL EXPENDITURE PROJECTS

Capital expenditure decisions, commonly known as *capital budgeting*, is the process of making long-term planning decisions for alternative investment opportunities. There are many investment decisions that the company may have to make in order to grow. Examples of capital budgeting applications are numerous. When should old facilities be replaced? Which piece of equipment should be selected to perform a particular operation? Should equipment be leased or purchased? Which product line should be selected? Should the company have one big plant or several smaller ones? Should the firm embark on a new research and development program?

WHAT ARE THE TYPES OF INVESTMENT PROJECTS?

There are typically two types of long-term investment decisions:

1. *Selection decisions* in terms of obtaining new facilities or expanding existing ones: Examples include:
   a. Investments in property, plant, and equipment as well as other types of assets.
   b. Resource commitments in the form of new product development, market research, introduction of a computer, refunding of long-term debt, and so on.
   c. Mergers and acquisitions in the form of buying another company to add a new product line.
2. *Replacement decisions* in terms of replacing existing facilities with new ones. Examples include replacing an old machine with a high-tech machine.

WHAT ARE THE FEATURES OF INVESTMENT PROJECTS?

Long-term investments have three important features:

1. They typically involve a large amount of initial cash outlay which tends to have a long-term impact on the firm's future profitability. Therefore, this initial cash outlay needs to be justified on a cost-benefit basis.
2. There are expected recurring cash inflows (for example, increased revenues, savings in cash operating expenses, etc.) over the life of the investment project. This frequently requires considering the *time value of money*.
3. Income taxes could make a difference in the accept or reject decision. Therefore, income tax factors must be taken into account in every capital budgeting decision.

HOW DO YOU MEASURE INVESTMENT WORTH?

Several methods of evaluating investment projects are as follows:

1. Payback period
2. Discounted payback period 3. Accounting (simple) rate of return (ARR)
4. Net present value (NPV)
5. Internal rate of return (IRR) (or time adjusted rate of return)
6. Profitability index (or present value index)
7. Equivalent uniform annual cost (UAC)

The NPV method and the IRR method are called *discounted cash flow (DCF) methods*. Each of these methods is discussed below.

Payback Period

The payback period measures the length of time required to recover the amount of initial investment. It is computed by dividing the initial investment by the cash inflows through increased revenues or cost savings.

Example 12.1. Assume:

Cost of investment	$18,000
Annual after-tax cash savings	$3,000

Then, the payback period is:

$$\text{Payback period} = \frac{\text{Initial investment}}{\text{Cost savings}} = \frac{\$18,000}{\$3,000} = 6 \text{ years}$$

Decision rule: Choose the project with the shorter payback period. The rationale behind this choice is: The shorter the payback period, the less risky the project, and the greater the liquidity.

Example 12.2. Consider two projects whose after-tax cash inflows are not even. Assume each project costs $1,000. The cash inflows are:

Year	A($)	B($)
1	100	500
2	200	400
3	300	300
4	400	100
5	500	
6	600	

When cash inflows are not even, the payback period has to be found by trial and error. The payback period of project A is ($1,000 = $100 + $200 + $300 + $400) over 4 years. The payback period of project B is $1,000 = $500 + $400 + $100):

$$2 \text{ years} + \frac{\$100}{\$300} = 2\,{}^1\!/_3 \text{ years}$$

Project B is the project of choice in this case, since it has the shorter payback period.

The advantages of using the payback period method of evaluating an investment project are that (1) it is simple to compute and easy to understand, and (2) it handles investment risk effectively.

The shortcomings of this method are that (1) it does not recognize the time value of money, and (2) it ignores the impact of cash inflows received after the payback period; essentially, cash flows after the payback period determine profitability of an investment.

Discounted Payback Period

You can take into account the time value of money by using the discounted payback period. The payback period will be longer using the discounted method since money is worth less over time. How do you determine the discounted payback period?

Discounted payback is computed by adding the present value of each year's cash inflows until they equal the initial investment.

$$\text{Discounted payback} = \frac{\text{Initial cash outlays}}{\text{Discounted annual cash inflows}}$$

Example 12.3. You invest $40,000 and receive the following cash inflows. The discounted payback period is calculated as follows:

Year	Cash inflows	T1 factor	Present value	Accumulated present value
1	$15,000	.9091	$13,637	$13,637
2	20,000	.8264	16,528	30,165
3	28,000	.7513	21,036	51,201

Thus,

$$\$30,165 + \frac{\$40,000 - 3,0165}{\$21,036} = 2 \text{ years} + .47 = 2.47 \text{ years}$$

Accounting (Simple) Rate of Return

Accounting rate of return (ARR) measures profitability from the conventional accounting standpoint by relating the required investment—or sometimes the average investment—to the future annual net income.

Decision rule: Under the ARR method, choose the project with the higher rate of return.

Example 12.4. Consider the following investment:

Initial investment	$6,500
Estimated life	20 years
Cash inflows per year	$1,000
Depreciation per year (using straight line)	$325

The accounting rate of return for this project is:

$$\text{ARR} = \frac{\text{Net income}}{\text{Investment}} = \frac{\$1,000 - \$325}{\$6,500} = 10.4\%$$

If average investment (usually assumed to be one-half of the original investment) is used, then:

$$ARR = \frac{\$1,000 - \$325}{\$3,250} = 20.8\%$$

The advantages of this method are that it is easily understood, simple to compute, and recognizes the profitability factor.

The shortcomings of this method are that it fails to recognize the time value of money, and it uses accounting data instead of cash flow data.

Net Present Value

Net present value (NPV) is the excess of the present value (PV) of cash inflows generated by the project over the amount of the initial investment (I):

$$NPV = PV - I$$

The present value of future cash flows is computed using the so-called cost of capital (or minimum required rate of return) as the discount rate. When cash inflows are uniform, the present value would be

$$PV = A \times T4 \ (i,n)$$

where A is the amount of the annuity. The value of T4 is found in Table A4, Appendix.

Decision rule: If NPV is positive, accept the project. Otherwise reject it.

Example 12.5. Consider the following investment:

Initial investment	$12,950
Estimated life	10 years
Annual cash inflows	$3,000
Cost of capital (minimum required rate of return)	12%

Present value of the cash inflows is:

$$PV = A.T4(i,n)$$
$$= \$3,000 \times T4(12\%,10 \text{ years})$$

$= \$3,000 \ (5.650)$	$16,950
Initial investment (I)	12,950
Net present value (NPV = PV − I)	$ 4,000

Since the NPV of the investment is positive, the investment should be accepted.

The advantages of the NPV method are that it obviously recognizes the time value of money and it is easy to compute whether the cash flows form an annuity or vary from period to period.

Internal Rate of Return

Internal rate of return (IRR), also called time adjusted rate of return, is defined as the rate of interest that equates I with the PV of future cash inflows. In other words,

at IRR, I = PV or NPV = 0

Decision rule: Accept the project if the IRR exceeds the cost of capital. Otherwise, reject it.

Example 12.6. Assume the same data given in Example 12.5, and set the following equality (I = PV):

$$\$12,950 = \$3,000 \times T4(i,10 \text{ years})$$

$$T4(i,10 \text{ years}) = \frac{\$12,950}{\$3,000} = 4.317$$

which stands somewhere between 18% and 20% in the 10-year line of Table A4 (Appendix). The interpolation follows:

PV of an annuity of $1 factor
T4(i,10 years)

18%	4.494	4.494
IRR	4.317	
20%		4.192
Difference	0.177	0.302

Therefore,

$$IRR = 18\% + \frac{0.177}{0.302}(20\% - 18\%)$$

$$= 18\% + 0.586(2\%) = 18\% + 1.17\% = 19.17\%$$

Since the IRR of the investment is greater than the cost of capital (12%), accept the project.

The advantage of using the IRR method is that it does consider the time value of money and therefore, is more exact and realistic than the ARR method.

The shortcomings of this method are that (1) it is time-consuming to compute, especially when the cash inflows are not even, although most financial calculators and PCs have a key to calculate IRR, and (2) it fails to recognize the varying sizes of investment in competing projects.

Profitability Index (or Present Value Index)

The profitability index is the ratio of the total PV of future cash inflows to the initial investment, that is, PV/I. This index is used as a means of ranking projects in descending order of attractiveness.

Decision rule: If the profitability index is greater than 1, then accept the project.

Example 12.7. Using the data in Example 12.5, the profitability index is

$$\frac{PV}{I} = \frac{\$16,950}{\$12,950} = 1.31$$

Since this project generates $1.31 for each dollar invested (i.e., its profitability index is greater than 1), accept the project.

The profitability index has the advantage of putting all projects on the same relative basis regardless of size.

Equivalent Uniform Annual Cost

In a facilities investment analysis, a comparison of annual costs requires the conversion of cash flows into an equivalent uniform annual series. This procedure may be thought of as the inverse of finding present values. The so-called *capital recovery factor*, which is the reciprocal of the present value of an annuity factor (T4), is used for this purpose. We calculate uniform annual cost (UAC) by using the following formula:

$$UAC = \{I - S_N [T3 (i,T)]\} CRF(i,N) + c$$

where

S_N = salvage value at terminal period N,
$CRT(i,N)$ = capital recovery factor = $1/T4$,
c   = uniform operating cost

Example 12.8. A company considers two equipment modules: Plan A and Plan B. Plan A, which requires an initial investment of $44,000, has expected annual operating costs of $24,400 and no terminal salvage value. Plan B, which requires an initial investment of $70,000, is expected reduce operating costs to $20,900 per year and have a salvage value of $9,500. Both equipment modules have a useful life of ten years. We assume that the firm has determined a cost of capital of 22%.

Using the UAC formula gives the following results:

$$UAC_A = \{\$44,000 - 0\ [T3\ (22\%,\ 10)]\}\ CRF(22\%,\ 10) + \$24,400$$
$$= \$44,000\ (1/T4(22\%,\ 10) + \$24,400 = \$44,000\ (1/3.923) + \$24,400$$
$$= \$44,000\ (0.255) + \$24,400 = \$35,620$$

$$UAC_B = \{\$70,000 - \$9,500[T3\ (22\%,\ 10)]\}\ CRF(22\%,\ 10) + \$24,400$$
$$= [\$70,000 - \$9,500(.137)](1/T4(22\%,\ 10) + \$20,900$$
$$= (\$70,000 - \$1,302)(1/3.293) + \$20,900$$
$$= \$68,698(0.255) + \$20,900 = \$38,418$$

Based on equivalent uniform annual cost comparison, Plan A is more economical than Plan B.

HOW TO SELECT THE BEST MIX OF PROJECTS WITH A LIMITED BUDGET

Many firms specify a limit on the overall budget for capital spending. Capital rationing is concerned with the problem of selecting the mix of acceptable projects that provides the highest overall NPV. The profitability index is used widely in ranking projects competing for limited funds.

Example 12.9. The Westmont Company has a fixed budget of $250,000. It needs to select a mix of acceptable projects from the following:

Projects	I($)	PV($)	NPV($)	Profitability Index	Ranking
A	70,000	112,000	42,000	1.6	1
B	100,000	145,000	45,000	1.45	2
C	110,000	126,500	16,500	1.15	5
D	60,000	79,000	19,000	1.32	3
E	40,000	38,000	−2,000	0.95	6
F	80,000	95,000	15,000	1.19	4

The ranking resulting from the profitability index shows that the company should select projects A, B, and D.

	I	PV
A	$70,000	$112,000
B	100,000	145,000
C	60,000	79,000
	$230,000	$336,000

Therefore,

$$NPV = \$336,000 - \$230,000 = \$106,000$$

HOW TO HANDLE MUTUALLY EXCLUSIVE INVESTMENTS

A project is said to be mutually exclusive if the acceptance of one project automatically excludes the acceptance of one or more other projects. In the case where one must choose between mutually exclusive investments, the NPV and IRR methods may result in contradictory indications. The conditions under which contradictory rankings can occur are:

1. Projects that have different life expectancies.
2. Projects that have different sizes of investment.
3. Projects whose cash flows differ over time. For example, the cash flows of one project increase over time, while those of another decrease.

The contradictions result from different assumptions with respect to the reinvestment rate on cash flows from the projects.

1. The NPV method discounts all cash flows at the cost of capital, thus implicitly assuming that these cash flows can be reinvested at this rate.
2. The IRR method implies a reinvestment rate at IRR. Thus, the implied reinvestment rate will differ from project to project.

The NPV method generally gives correct ranking, since the cost of capital is a more realistic reinvestment rate.

Example 12.10. Assume the following cash flows:

	0	1	2	3	4	5
A	(100)	120				
B	(100)					201.14

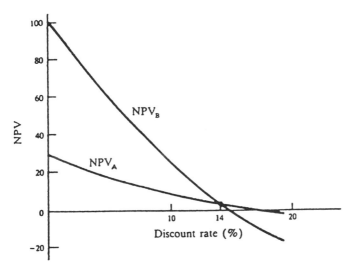

Figure 12.1 NPV profile.

Computing IRR and NPV at 10% gives the following different rankings:

	IRR	NPV at 10%
A	20%	9.01
B	15%	24.90

THE NPVs plotted against the appropriate discount rates form a graph called a NPV profile (Figure 12.1).

At a discount rate larger than 14 percent, A has a higher NPV than B. Therefore, A should be selected. At a discount rate less than 14 percent, B has a higher NPV than A, and thus should be selected. The correct decision is to select the project with the higher NPV, since the NPV method assumes a more realistic reinvestment rate, that is, the cost of capital.

LEASE-PURCHASE DECISION

The lease-purchase decision is one that commonly confronts firms considering the acquisition of new assets. It is a hybrid capital budgeting decision which forces a company to compare the leasing and purchasing alternatives. To make an intelligent decision, an after-tax, cash outflow, present value comparison is needed. There are special steps to take when making this comparison.

When considering a lease, take the following steps:

1. Find the annual lease payment. Since the annual lease payment is typically made in advance, the formula used is:

$$\text{Amount of lease} = A + A \times T4(i,n-1) \text{ or } A = \frac{\text{Amount of lease}}{1 + T4(i,n-1)}$$

Notice we use n−1 rather than n.

2. Find the after-tax cash outflows.
3. Find the present value of the after-tax cash outflows.

When considering a purchase, take the following steps:

1. Find the annual loan amortization by using:

$$A = \frac{\text{Amount of loan for the purchase}}{T4(i,n-1)}$$

This step may not be necessary since this amount is usually available.
2. Calculate the interest. The interest is segregated from the principal in each of the annual loan payments because only the interest is tax-deductible.
3. Find the cash outflows by adding interest and depreciation (plus any maintenance costs), and then compute the after-tax outflows.
4. Find the present value of the after-tax cash outflows, using Table A3, Appendix.

Example 12.11. A firm has decided to acquire an asset costing $100,000 that has an expected life of 5 years, after which the asset is not expected to have any residual value. The asset can be purchased by borrowing or it can be leased. If leasing is used, the lessor requires a 12% return. As is customary, lease payments are to be made in advance, that is, at the end of the year prior to each of the 10 years. The tax rate is 50% and the firm's cost of capital, or after-tax cost of borrowing, is 8%.

First, compute the present value of the after-tax cash outflows associated with the leasing alternative.

1. Find the annual lease payment:

$$A = \frac{\text{Amount of lease}}{1 + T4(i,n-1)}$$

Year	(1) Lease payment($)	(2) Tax savings($)	(3) = (1) − (2) After-tax cash outflow($)	(4) PV at 8%	(5)=(3) × (4) PV of cash out-flow($,rounded)
0	23,216		23,216	1.000	23,216
1-4	23,216	11,608[a]	11,608	3.3121[b]	38,447
5		11,608	(11,608)	0.6806[c]	(7,900)
					53,763

[a]$23,216 × 50% [b]From Table 4. [c]From Table 3.

$$= \frac{\$100,000}{1 + T4(12\%,4 \text{ years})} = \frac{\$100,000}{1 + 3.3073} = \frac{\$100,000}{4.3073} = \$23,216 \text{ (rounded)}$$

Steps 2 and 3 can be done in the same schedule, as illustrated.

If the asset is purchased, the firm is assumed to finance it entirely with a 10% unsecured term loan. Straight-line depreciation is used with no salvage value. Therefore, the annual depreciation is $20,000 ($100,000/5 years). In this alternative, first find the annual loan payment (1) by using:

$$A = \frac{\text{Amount of loan}}{T4(i,n)}$$

$$A = \frac{\$100,000}{T4(10\%,5 \text{ years})} = \frac{\$100,000}{3.7906} = \$26,381 \text{ (rounded)}$$

2. Calculate the interest by setting up a loan amortization schedule.

Yr.	(1) Loan payment($)	(2) Beginning-of-yr. principal($)	(3)=(2)(10%) Interest($)	(4)=(1)-(3) Principal($)	(5)=(2)-(4) End-of-yr. principal
1	26,381	100,000	10,000	16,381	83,619
2	26,381	83,619	8,362	18,019	65,600
3	26,381	65,600	6,560	19,821	45,779
4	26,381	45,779	4,578	21,803	23,976
5	26,381	23,976[a]	2,398	23,983[a]	

[a]Because of rounding errors, there is a slight difference between (2) and (4)

Steps 3 (cash outflows) and 4 (present values of those outflows) can be calculated as shown in Figure 12.2.

Year	(1) Loan Payment ($)	(2) Interest ($)	(3) Depreciation ($)	(4) = (2) + (3) Total Deductions ($)	(5) = (4)(50%) Tax Savings ($)	(6) = (1) − (5) Cash Outflow($)	(7) PV at 8%	(8) = (6) × (7) PV of Cash Outflow ($)
1	26,381	10,000	20,000	30,000	15,000	11,381	0.9259	10,538
2	26,381	8,362	20,000	28,362	14,181	12,200	0.8573	10,459
3	26,381	6,560	20,000	26,560	13,280	13,101	0.7938	10,400
4	26,381	4,578	20,000	24,578	12,289	14,092	0.7350	10,358
5	26,381	2,398	20,000	22,398	11,199	15,182	0.6806	10,333
								52,088

Figure 12.2 Present value calculations.

The sum of the present values of the cash outflows for leasing and purchasing by borrowing shows that purchasing is preferable because the PV of borrowing is less than the PV of leasing ($52,008 versus $53,763). The incremental savings is $1,675.

HOW DO INCOME TAXES AFFECT INVESTMENT DECISIONS?

Income taxes make a difference in many capital budgeting decisions. In other words, the project which is attractive on a before-tax basis may have to be rejected on an after-tax basis. Income taxes typically affect both the amount and the timing of cash flows. Since net income, not cash inflows, is subject to tax, after-tax cash inflows are not usually the same as after-tax net income.

How to Calculate After-Tax Cash Flows

Let us define:

S = Sales
E = Cash operating expenses
d = Depreciation
t = Tax rate

Then, before-tax cash inflows (or cash savings) = S − E and net income
$$= S - E - d$$

By definition,
After-tax cash inflows = Before-tax cash inflows − Taxes
$$= (S - E) - (S - E - d)(t)$$

Rearranging gives the short-cut formula:
After-tax cash inflows = (S − E)(1 − t) + (d)(t) or
$$= (S - E - d) (1 - t) + d$$

As can be seen, the deductibility of depreciation from sales in arriving at taxable net income reduces income tax payments and thus serves as a *tax shield*.

Tax shield = Tax savings on depreciation = (d)(t)

Example 12.12. Assume:

S = $12,000
E = $10,000
d = $500 per year using the straight line method
t = 30%

Then,

$$\text{After-tax cash inflow} = (\$12,000 - \$10,000)(1 - .3) + (\$500)(.3)$$
$$= (\$2,000)(.7) + (\$500)(.3)$$
$$= \$1,400 + \$150 = \$1,550$$

Note that a tax shield = tax savings on depreciation = (d)(t)

$$= (\$500)(.3) = \$150$$

Since the tax shield is dt, the higher the depreciation deduction, the higher the tax savings on depreciation. Therefore, an accelerated depreciation method (such as double-declining balance) produces higher tax savings than the straight-line method. Accelerated methods produce higher present values for the tax savings which may make a given investment more attractive.

Example 12.13. The Shalimar Company estimates that it can save $2,500 a year in cash operating costs for the next ten years if it buys a special-purpose machine at a cost of $10,000. No residual value is expected. Depreciation is by the straight-line method. Assume that the income tax rate is 30%, and the after-tax cost of capital (minimum required rate of return) is 10%. After-tax cash savings can be calculated as follows:

Note that depreciation by straight-line is $10,000/10 = $1,000 per year. Thus,

$$\text{After-tax cash savings} = (S - E)(1 - t) + (d)(t)$$
$$= \$2,500(1 - .3) + \$1,000(.3)$$
$$= \$1,750 + \$300 = \$2,050$$

To see if this machine should be purchased, the net present value can be calculated.

PV = $2,050 T4(10%, 10 years) = $2,050 (6.145)=$12,597.25

Thus, NPV = PV − I = $12,597.25 − $10,000 = $2,597.25

Since NPV is positive, the machine should be bought.

Example 12.14. XYZ Corporation has provided its revenues and cash operating costs (excluding depreciation) for the old and the new machines, as follows:

	Revenue	Annual cash operating costs	Net profit before depreciation and taxes
Old machine	$150,000	$70,000	$80,000
New machine	$180,000	$60,000	$120,000

Assume that the annual depreciation of the old machine and the new machine will be $30,000 and $50,000, respectively. Assume further that the tax rate is 46%.

To arrive at net profit after taxes, we must first deduct depreciation expense from the net profit before depreciation and taxes, as follows:

	Net Profits after Taxes	Add depreciation	After-tax cash inflows
Old machine	($80,000 − $30,000)(1 − 0.46) = $27,000	$30,000	$57,000
New machine	($120,000 − $50,000)(1 − 0.46) = $37,800	$50,000	$87,800

Subtracting the after-tax cash inflows of the old machine from the cash inflows of the new machine results in the relevant, or incremental, cash inflows for each year.

Therefore, in this example, the relevant or incremental cash inflows for each year are $87,800 − $57,000 = $30,800.

Alternatively, the incremental cash inflows after taxes can be computed, using the following simple formula:

After-tax incremental cash inflows = (increase in revenues)(1 − tax rate) − (increase in cash charges)(1 − tax rate) + (increase in depreciation expenses)(tax rate)

Example 12.15. Using the data in Example 12.14, after-tax incremental cash inflows for each year are:

Increase in revenue × (1 − tax rate):	
($180,000 − $150,000)(1 − 0.46)	$16,200
− Increase in cash charges × (1 − tax rate):	
($60,000 − $70,000)(1 − 0.46)	−(−5,400)
+ Increase in depreciation expense × tax rate:	
($50,000 − $30,000)(0.46)	9,200
	$30,800

HOW DOES MACRS AFFECT INVESTMENT DECISIONS?

We saw that depreciation provided the tax shield in the form of dt. Among the commonly used depreciation methods are straight-line and accelerated methods such as sum-of-the years'-digits (SYD) and double-declining-balance (DDB), which was discussed in an earlier chapter. The following should be noted:

1. Over the life of the fixed asset, the total depreciation charge will be the same no matter what depreciation method is used; only the timing of the tax savings will differ.
2. The accelerated methods such as SYD and DDB are advantageous for tax purposes since higher depreciation charges in the earlier years result in less income and thus less taxes. The tax savings may then be invested for a return.

Although the traditional depreciation methods can still be used for computing depreciation for book purposes, 1981 saw a new way of computing depreciation deductions for tax purposes. The current rule is called the *Modified Accelerated Cost Recovery System* (MACRS) rule, as enacted by Congress in 1981 and then modified somewhat in 1986 under the Tax Reform Act of 1986. This rule is characterized as follows:

1. It abandons the concept of useful life and accelerates depreciation deductions by placing all depreciable assets into one of eight age property classes. It calculates deductions, based on an allowable percentage of the asset's original cost (see Tables 12.1 and 12.2).

   With a shorter life than useful life, the company would be able to deduct depreciation more quickly and save more in income taxes in the earlier years, thereby making an investment more attractive. The rationale behind the system is that in this way the government encourages the company to invest in facilities and increase its productive capacity and efficiency. (Remember that the higher d, the larger the tax shield dt).
2. Since the allowable percentages in Table 12.2 add up to 100%, there is no need to consider the salvage value of an asset in computing depreciation.
3. The company may elect the straight-line method. The straight-line convention must follow what is called the *half-year convention*. This means that the company can deduct only half of the regular straight-line depreciation amount in the first year. The reason for electing to use the MACRS optional straight-line method is that some firms may prefer to stretch out depreciation deductions using the straight-line method rather than to accelerate them. Those firms are the ones that just start out or has little or no income and wish to show more income on their income statements.

Example 12.16. Assume that a machine falls under a 3-year property class and costs $3,000 initially. The straight-line option under MACRS differs from the traditional straight-line method in that under this method the company would deduct only $500 depreciation in the first year and the fourth year ($3,000/3 years = $1,000; $1,000/2 = $500). The table compares the straight-line with half-year convention with the MACRS.

Year	Straight-line (half-year) depreciation	Cost		MACRS %	MACRS deduction
1	$ 500	$3,000	×	33.3%	$ 999
2	1,000	3,000	×	44.5	1,335
3	1,000	3,000	×	14.8	444
4	500	3,000	×	7.4	222
	$3,000				$3,000

Table 12.1 Modified Accelerated Cost Recovery System Classification of Assets

Year	Property class					
	3-year	5-year	7-year	10-year	15-year	20-year
1	33.3%	20.0%	14.3%	10.0%	5.0%	3.8%
2	44.5	32.0	24.5	18.0	9.5	7.2
3	14.8[a]	19.2	17.5	14.4	8.6	6.7
4	7.4	11.5[a]	12.5	11.5	7.7	6.2
5		11.5	8.9[a]	9.2	6.9	5.7
6		5.8	8.9	7.4	6.2	5.3
7			8.9	6.6[a]	5.9[a]	4.9
8			4.5	6.6	5.9	4.5[a]
9				6.5	5.9	4.5
10				6.5	5.9	4.5
11				3.3	5.9	4.5
12					5.9	4.5
13					5.9	4.5
14					5.9	4.5
15					5.9	4.5
16					3.0	4.4
17						4.4
18						4.4
19						4.4
20						4.4
21						2.2
Total	100%	100%	100%	100%	100%	100%

[a]Denotes the year of changeover to straight-line depreciation.

Table 12.2 MACRS Tables by Property Class

ACRS property class and depreciation method	Useful life (ADR midpoint life)[a]	Examples of assets
3-year property 200% declining balance	4 years or less	Most small tools are included; the law specifically excludes autos and light trucks from this property class.
5-year property 200% declining	More than 4 years to less than 10 years	Autos and light trucks, computers, typewriters, balance copiers, duplicating, equipment, heavy general-purpose trucks, and research and experimentation equipment are included.
7-year property 200% declining	10 years or more to less than 16 years	Office furniture and fixtures and most items balance of machinery and equipment used in production are included.
10-year property 200% declining balance	16 years or more to less than 20 years	Various machinery and equipment, such as that used in petroleum distilling and refining and in the milling of grain, are included.
15-year property 150% declining balance	20 years or more to less than 25 years	Sewage treatment plants telephone and electrical distribution facilities, and land improvements are included.
20-year property 150% declining balance	25 years or more	Service stations and other real property with an ADR midpoint life of less than 27.5 years are included.
27.5-year property Straight-line	Not applicable	All residential rental property is included.
31.5-year property Straight-line	Not applicable	All nonresidential real property is included.

[a]The term ADR midpoint life means the "useful life" of an asset in a business sense; the appropriate ADR midpoint lives for assets are designated in the tax regulations.

	Present value factor @ 10%	Present value
Initial investment: $10,000	1.000	$(10,000.00)
(S − E)(1 − t):		
$5,000 (1 − .3) = $3,500 for 4 years	3.170[a]	$11,095.00

(d)(t):

Year	Cost	MACRS %	d	(d)(t)		
1	$10,000 ×	33.3%	$3,330	$ 999	.909[b]	908.09
2	$10,000 ×	44.5	4,450	1,335	.826[b]	1,102.71
3	$10,000 ×	14.8	1,480	444	.751[b]	333.44
4	$10,000 ×	7.4	740	222	.683[b]	151.63

Salvage value:

$1,200 in year 4: $1,200 (1 − .3) =		
$840[c]	.683[b]	573.72
Net present value (NPV)		$4,164.59

[a]T4(10%, 4 years) = 3.170 (from Table 4, Appendix).
[b]T3 values obtained from Table A3, Appendix.
[c]Any salvage value received under the MACRS rules is a *taxable gain* (the excess of the selling price over book value, $1,200 in this example), since the book value will be zero at the end of the life of the machine.

Example 12.17. A machine costs $10,000. Annual cash inflows are expected to be $5,000. The machine will be depreciated using the MACRS rule and will fall under the 3-year property class. The cost of capital after taxes is 10%. The estimated life of the machine is 4 years. The salvage value of the machine at the end of the fourth year is expected to be $1,200. The tax rate is 30%.

The formula for computation of after-tax cash inflows (S − E)(1 − t) + (d)(t) needs to be computed separately. The NPV analysis can be performed as shown in the table.

Since, as described in the table, NPV = PV - I = $4,164.59 is positive, the machine should be bought.

Tax Effects of Disposal

In general, gains and losses on disposal of equipment are taxed in the same way as ordinary gains and losses. Immediate disposal of the old equipment results in a loss that is fully tax deductible from current income. The loss (the excess of the book value over the disposal value) must be computed to isolate its effect on current income tax, but the total cash inflow is the selling price *plus* the current income tax benefit.

Example 12.18. Assume that the equipment has a salvage value of $1,200, while its book (undepreciated) value is $2,000. *Two* cash inflows are connected with this sale. The tax rate is 30%.

1. A $1,200 cash inflow in the form of the sale price, and
2. A $240 cash inflow in the form of a reduction in income taxes, resulting from the tax shield provided by the loss sustained on the sale, just like the tax shield provided by depreciation deduction, as computed as follows:

Book value	$2,000	
Selling price	1,200	
Loss	800	
Tax shield	× .3	$240

Thus, the total cash inflow from the disposal is $1,440 ($1,200 + $240).

WHAT TO KNOW ABOUT THE COST OF CAPITAL

The cost of capital is defined as the rate of return that is necessary to maintain the market value of the firm (or price of the firm's stock). Project managers must know if the cost of capital, often called the *minimum required rate of return*, was used either as a discount rate under the NPV method or as a hurdle rate under the IRR method used earlier in the chapter and also used in calculating the residual income (RI) in Chapter 9. The cost of capital is computed as a weighted average of the various capital components, which are the items on the right-hand side of the balance sheet such as debt, preferred stock, common stock, and retained earnings.

Cost of Debt

The cost of debt is stated on an after-tax basis, since the interest on the debt is tax deductible. However, the cost of preferred stock is the stated annual dividend rate. This rate is not adjusted for income taxes because the preferred dividend, unlike debt interest, is not a deductible expense in computing corporate income taxes.

Example 12.19. Assume that the Hume Company issues a $1,000, 8%, 20-year bond whose net proceeds are $940. The tax rate is 40%. Then the after-tax cost of debt is:

$$= 8.00\% \, (1 - 0.4) = 4.8\%$$

Example 12.20. Suppose that the Hume company has preferred stock that pays a $12 dividend per share and sells for $100 per share in the market. Then the cost of preferred stock is:

$$\frac{\text{Dividend per share}}{\text{Price per share}} = \frac{\$12}{\$100} = 12\%$$

Cost of Common Stock

The cost of common stock is generally viewed as the rate of return investors require on a firm's common stock. One way to measure the cost of common stock is to use the *Gordon's growth model*. The model is

$$P_0 = \frac{D_1}{r - g}$$

where

P_0 = Value (or market price) of common stock
D_1 = Dividend to be received in 1 year
r = Investor's required rate of return
g = Rate of growth (assumed to be constant over time)

Solving the model for r results in the formula for the cost of common stock:

$$r = \frac{D_1}{P_0} + g$$

Example 12.21. Assume that the market price of the Hume Company's stock is $40. The dividend to be paid at the end of the coming year is $4 per share and is expected to grow at a constant annual rate of 6%. Then the cost of this common stock is:

$$\frac{D_1}{P_0} + g = \frac{\$4}{\$40} + 6\% = 16\%$$

Cost of Retained Earnings

The cost of retained earnings, ks, is closely related to the cost of existing common stock, since the cost of equity obtained by retained earnings is the same as the rate of return investors require on the firm's common stock.

Measuring the Overall Cost of Capital

The firm's overall cost of capital is the weighted average of the individual capital costs, with the weights being the proportions of each type of capital used.

Σ (percentage of the total capital structure supplied by each source of capital × cost of capital for each source)

The computation of overall cost of capital is illustrated in the following example.

Example 12.22. Assume that the capital structure at the latest statement date is indicative of the proportions of financing that the company intends to use over time (the cost percent is given at the right):

Mortgage bonds ($1,000 par)	$20,000,000	4.80% (from Example 12.17)
Preferred stock ($100 par)	5,000,000	12.00 (from Example 12.18)
Common stock ($40 par)	20,000,000	16.00 (from Example 12.19)
Retained earnings	5,000,000	16.00%
Total	$50,000,000	

These proportions would be applied to the assumed individual explicit after-tax costs shown in the small table.
The overall cost of capital is 11.12%

Source	Weights	Cost	Weighted cost
Debt	40%[a]	4.80%	1.92%[b]
Preferred stock	10	12.00%	1.20
Common stock	40	16.00%	6.40
Retained earnings	10	16.00%	1.60
	100%		11.12%

[a]$20,000,000/$50,000,000 = .40 = 40%
[b]4.80% × 40% = 1.92%

By computing a company's cost of capital, we can determine its minimum rate of return, which is used as the discount rate in present value calculations and in calculating an investment center's residual income (RI). A company's cost of capital is also an indicator of risk. For example, if your company's cost of financing increases, it is being viewed as more risky by investors and creditors, who are demanding a higher return on their investments in the form of higher dividend and interest rates.

CONCLUSION

We have examined the process of evaluating investment projects. We have also discussed five commonly used criteria for evaluating capital budgeting projects, including the net present value (NPV) and internal rate of return (IRR) methods. The problems that arise with mutually exclusive investments and capital rationing were addressed.

Since income taxes could make a difference in the accept or reject decision, tax factors must be taken into account in every decision.

Although the traditional depreciation methods can still be used for computing depreciation for book purposes, 1981 saw a new way of computing depreciation deductions for tax purposes. The rule is called the modified accelerated cost recovery system (MACRS). It was enacted by Congress in 1981 and then modified somewhat in 1986 under the Tax Reform Act of 1986. We presented an overview of the traditional depreciation methods and illustrated the use of MACRS.

It is important to understand how to calculate a firm's cost of capital. The cost of capital, which is the minimum required rate of return that maintains the firm's market value, is used as the discount rate in calculating the net present value (NPV) of a project and the hurdle rate (cutoff) rate for the IRR and RI calculations. The cost of capital is essentially a firm's weighted average of specific costs of capital.

13

HOW TAXES AFFECT BUSINESS DECISIONS

A corporation is recognized as a separate legal and taxable entity, and it pays its own taxes to many federal, state, and local taxing agencies. If the corporation is operating internationally, it has the additional problem of complying with the tax laws of foreign countries. Taxes may be levied on income, sales, and property. The federal corporate income tax (filed on Form 1120) is the most important because it often represents the largest tax liability. Thus, it can have a major effect on your financial decisions.

To make sound financial and investment decisions, you must have an understanding of the basic concepts underlying the U.S. tax structure and how they affect your decisions.

This chapter first discusses the general structure of the corporate income tax, such as taxable income, capital gains and losses, deductible expenses, net operating loss carrybacks and carryforwards, tax rates, and tax prepayments and credits. Then, it reviews various tax planning strategies to help minimize your company's income tax liability in the current year and postpone the payment of taxes to later years. The advantages of electing S corporation status will also be discussed. *NOTE*: The tax rules discussed in this chapter have been updated for changes brought about by the Revenue Reconciliation Act of 1993.

The purpose of this chapter is to provide a brief introduction to corporate income taxation and help the reader (engineer) develop a framework for understanding the pervasive effects of taxation on business decisions. For clarity of presentation, only important concepts and general tax rules will be emphasized, leaving out many complex details and special applications. The reader should be aware that tax laws must necessarily be very complex to deal with the myriad of possible situations in actual practice. Also, since tax laws are man made and are

---

This chapter was written by Loc T. Nguyen, LL.M., CPA, a tax consultant to industry and Professor of Taxation at California State University, Long Beach, CA.

created by Congress, the tax rules often have many exceptions and exceptions to the exceptions to accommodate special situations or special interest groups.

THE CORPORATE INCOME TAX

A regular corporation is subject to double taxation. It is taxed first as a separate entity at the corporate level on the corporation's taxable income. Then, when the after-tax profits are distributed as dividends to the shareholders, they are again taxed at the shareholders' level.

The general tax formula for corporations is as follows:

Gross income
Less: Deductions
= Taxable income
× Tax rates
= Gross tax
Less: Tax credits and prepayments
= Tax due (or Refund)

A brief review of the various elements of this tax formula will give us the general understanding needed for effective tax planning.

Gross Income

Corporate gross income is of two types:

1. *Ordinary income*, which includes income from normal business operations as well as miscellaneous incomes such as rents, interest, and dividends. Ordinary income is fully taxable, however the corporation has a special exclusion for dividends received from taxable domestic corporations. Generally, a regular corporation that owns less than 20% of another domestic corporation from which it receives dividends is entitled to deduct 70% of the dividends received. For example, if a $100 dividend is received by the corporation, $70 can be excluded and only $30 of the dividend is included in taxable income. This 70% exclusion can be increased to 80% if the corporation receiving the dividend owns at least 20% but less than 80% of the dividend paying corporation's stock. And for ownership exceeding 80% of the dividend paying corporation, the dividend received deduction will be 100%. The reason for this special corporate deduction is the double tax already imposed on the corporation's taxable income. Without the dividends received deduction, dividends received would be subject to triple taxation, and corporate ownership of another corporation's stock would be prohibitively expensive taxwise.

2. *Capital gains or losses*, resulting from the sale of capital assets. In general, assets bought and sold in the ordinary course of a company's business, such as inventory, receivables, and depreciable assets, generate ordinary income. The sale or exchange of all other assets, such as stocks, bonds, real estate investments, generate capital gains or losses.

Capital gains and losses are netted together and any net capital gain is taxed at the same rate as ordinary income, up to a maximum tax rate of 35%. However, unlike the case with individuals, corporations cannot deduct a net capital loss from ordinary income in the year of the loss, i.e. there is no capital loss deduction. Thus, a corporation's capital losses can only be used to offset the current year capital gains. However, any unused capital loss may be carried back to the three years preceding the year of loss to offset capital gains and, if not completely absorbed, may be then carried forward for up to five years subsequent to the loss year.

Example 13.1. If a corporation has a net capital loss in 1994, the loss may be carried back to 1991 to offset any capital gain reported by the corporation in that year. Any remaining unused capital loss is then carried to 1992 and, if necessary, to 1993. After carrying back the current year capital loss to the three prior years, the corporation can continue to carry forward any remaining capital loss to offset capital gains realized in the next five years, 1995 to 1999. Any loss not utilized by the end of 1999 will expire without tax benefit.

Capital gains and losses are classified as long-term or short-term. The gain is short-term if the holding period of the asset is one year or less. When the holding period exceeds one year, the gain is considered long-term. Whether the capital gain is short-term or long-term, it is taxed as ordinary income, up to a maximum tax rate of 35%.

Example 13.2. A company sells securities it owns resulting in a capital gain of $180,000. The tax on the capital gain assuming the company is in the maximum tax bracket is:

$$\$180,000 \times 35\% = \$63,400.$$

Deductions

All of a regular corporation's activities are considered to be connected with the conduct of a trade or business. Therefore, all the ordinary and necessary expenses paid or incurred during the year by the business are tax deductible.

In 1994, the deduction for business meals and entertainment expenses has been reduced from 80% to 50%. The meal must have a direct relationship to carrying out your business activities. Business must be discussed before, during, and after the meal. "Quiet meals", where no business is discussed before, during, or after the meals, are no longer deductible. The meal or beverage cannot be "lavish or extravagant" under the circumstances. Transportation to and from the restaurant is still 100% deductible. However, parking at a sports arena is only 50% deductible.

Example 13.3. After business discussions in the office, you took a customer out to dinner and drinks, and it cost $125. The business meal is limited in terms of tax deductibility to $120 \times 50\% = \$60$. In addition, the taxi fare to the restaurant was $20. This is fully deductible.

Generally, food- and entertainment- related employee benefits are fully deductible. Examples are departmental and office parties. However, the deduction for club dues has been eliminated. This applies to all types of clubs, including those for business, social, athletic, luncheon, sporting, airline, and hotel clubs. Lobbying expenses, i.e. expenses related to attempts to influence federal and state legislators, are no longer deductible as ordinary and necessary business expenses.

Promotional items for public distribution are fully deductible (e.g. samples). Deductions for business gifts are limited to $25 per individual customer, client, or other business contact. Items clearly of an advertising nature which costs $4 or less and are imprinted with the company's name (e.g. pens, calendars, etc...) do not have to be included in the $25 limitation.

Example 13.4. You gave business gifts to 17 customers. These gifts, which were not of an advertising nature, had the following fair market value: 4 @ $10, 4 @ $25, 4 @ $50, and 5 @ $100. These gifts are deductible as business expenses for $365 computed as follows:

4 @ $10	$ 40
13 @ $25	325
Total	$365

All bad debts incurred by the corporation are business-related and can be used to offset the corporation's ordinary income. The corporation can no longer use the reserve method which allows deductions for estimated bad debts. Instead, it must use the direct write-off method which allows a bad debt deduction only at the time the debt actually becomes partially or totally worthless.

A charitable contribution made in cash or property is deductible. Gifts of appreciated long-term capital gain property are deductible at the full fair market value of the donated property. For contributions of inventory used to care for the ill, the needy, or infants and contributions of scientific property to an educational institution to be used for research, the amount of deduction is the cost of the property plus one half of the potential profit (but not exceeding two times the cost). However, gifts of inventory and short-term capital gain property used in a manner not related to the organization's exempt purpose, are deductible only at cost. The deduction for charitable contributions is generally limited to 10% of taxable income, computed by excluding the charitable contribution itself, the dividends received exclusion, and the net operating loss and capital loss carry-backs. Any charitable contribution in excess of the limitation may be carried forward and deducted in any of the five succeeding years.

With life insurance premiums, there are some circumstances in which a corporation may claim a deduction. These include premiums paid for "key man" life insurance (i.e. life insurance for any officer, employee who may have a financial interest in the company) and for group-term life insurance when the company is not the beneficiary. "Key man" life insurance premiums are deductible only if (1) the premium payments are an ordinary business expense as additional compensation to the "key" employee, and (2) the total amount of all compensation paid to that employee is not unreasonable.

As for group-term insurance, the premiums paid for coverage of up to $50,000 per employee are a deductible expense for the corporation and are not taxable income to the employee. For coverage above $50,000, the employee must include in income an amount equivalent to the cost of the additional protection. The group-term insurance policy must be non-discriminatory, i.e. it must provide coverage for all employees with only few permitted exceptions such as those who work part-time, who are under age 21, or who have not been employed for at least 6 months.

Special Tax Incentives for Businesses Located in Economically Disadvantaged Areas

There are now nine "empowerment zones" and up to 95 "enterprise communities" located in economically disadvantaged areas. Businesses located in these areas qualify for special tax incentives, including a 20% wage credit for the first $15,000 of wages paid to a zone resident working in the zone and increased expensing for depreciable business property up to a total of $37,500.

Other deductible expenses include interest, professional fees (e.g. independent accountants, outside attorneys), casualty and theft losses, and bad debts when a specific amount is deemed uncollectible. Nondeductible expenses include fines, penalties, and illegal payments.

Depreciation

A business buying machinery, equipment, buildings, or other income-producing property expected to last more than one year cannot deduct the entire cost in the

year of purchase. If the property or asset is depreciable, the cost must be capital-
ized and recovered through depreciation allowances over its useful life.

The corporation has the choice of depreciating tangible assets under the
straight-line method or the modified accelerated cost recovery (MACRS)
method. The basis for determining depreciation is the cost of the assets and
improvements. The depreciation method for tangible property is 200% declining
balance with a later change to straight-line.

If land and building are purchased together, the purchase price must be
apportioned between the land and the building because only the building cost is
depreciable. The portion allocated to the land cannot be depreciated since land is
considered to have unlimited useful life. The building must be depreciated using
the straight-line method over 27.5 years if used for residential purposes and over
39 years (increased from 31.5 years) if used for non-residential purposes. In the
year of acquisition (or disposition), the mid-month convention applies to real
property, i.e. the property is treated as placed in service (or disposed of) in the
middle of the month.

All personal property is subject to the half-year convention, i.e. property is
treated as placed in service (or disposed of) in the middle of the year. However,
if more than 40% of the assets purchased during the year are acquired in the last
quarter of the year, then the company is required to use the mid-quarter convention and apply it to all personal property placed in service or disposed of during
the year.

Example 13.5. XYZ Co. places two items of five-year property in service in
1995. Machine A costs $59,000 and machine B costs $41,000. If the half-year
convention applies, the XYZ cost recovery deduction is $20,000, computed as
follows:

($59,000 + $41,000) : 5 years × 200%DB × .50 = $20,000

If machine A is placed in service in January and machine B is placed in service
in December, then the mid-quarter convention applies since more than 40% of
the assets purchased during the year was purchased in the last quarter. The
depreciation is then computed as follows:

Machine A: $59,000 : 5 years × 200%DB × (10.5 : 12) = $20,650
Machine B: $41,000 : 5 years × 200%DB × (1.5 : 12) =   2,050
Total depreciation: $22,700

MACRS depreciation lowers taxes in the early years of an asset's life, thus
improving corporate cash flow and making funds available for investment. The
faster the cash flow, the greater the rate of return earned on the investment.

Your company may elect to immediately deduct part or all of the cost of a depreciable asset in the year of purchase rather than depreciating it over its useful life. This provision relates to personal property qualifying under MACRS and acquired for trade or business use.

The immediate expensing election is not applicable to real estate. The maximum amount that may be expensed has been increased from $10,000 to $17,500 per year. However, if the purchase of depreciable equipment exceeds $200,000, the $17,500 immediate expensing amount is reduced at the rate of $1.00 for each dollar in excess of $200,000. Accordingly, no immediate expensing is allowed if total purchases of assets during the year exceed $217,500.

Furthermore, the amount expensed for any year may not exceed the taxable income of the corporation for that year. Any disallowed amount may be carried forward and deductible in succeeding taxable years.

Example 13.6. In 1995, ABC Corporation purchases and places in service $205,000 of qualifying equipment. The taxable income for the corporation (before the immediate expensing deduction) is $10,000. The $17,500 ceiling amount for immediate expensing is initially reduced by $5,000 ($205,000 − $200,000). The remaining balance of $12,500 is further reduced by $2,500 because of the taxable income limitation. ABC's immediate expensing is $10,000 in 1995, with $2,500 carryover to 1996. The cost basis of the equipment is reduced by $12,500 in 1995. The remaining balance of $192,500 ($205,000 − $12,500) of the assets is subject to the regular MACRS depreciation:

$192,500 : 5 years x 200%DB × .50 = $38,500

Total depreciation for 1995 is $48,500 ($10,000 + $38,500), with $2,500 carryover to 1996.

Business automobiles are subject to a special set of depreciation rules. Starting in 1987, automobiles must be depreciated over five years. In addition, the maximum amount of annual depreciation in 1994 is subject to the following ceilings:

1st year	$2,860
2nd year	4,600
3rd year	2,750
4th year and thereafter	1,675

The effect of these rules is to limit the annual depreciation for cars costing more than $13,560. For the more expensive cars, depreciation can be continued after the fifth year, subject to the maximum annual limitation. In addition, if the car is used less than 100% for business, the maximum amount of annual depreciation allowed must be reduced proportionately.

Depreciation and how it affects long-term capital investment decisions was fully discussed in Chapter 12.

Amortization

Some types of intangible assets may be amortized (i.e. written off systematically over their estimated useful lives) for tax purposes. For instance, a new company's organization costs may be amortized over a period of not less than 60 months. Goodwill and going concern value, which used to be non-amortizable, can now be depreciated or amortized on a straight-line basis over a 15 year period.

Passive Activity Losses

In general, losses attributable to a passive activity (e.g. a limited partnership interest or rental activities) can only be offset against passive activity income, and any net passive losses are not deductible against ordinary income.

These passive losses must be carried forward to future years and they can only be deducted in full in the year the asset is sold. The limitation on passive activity losses applies to all individual taxpayers as well as to closely held and personal service corporations. However, a new exception to this rule is that losses from rental real estate are not subject to the general limitation for closely-held corporations engaged in real property trades and businesses.

The corporation qualifies for this exception when more than 50% of the corporation's gross receipts for the year are from real property development, construction, rental operation, management, leasing, or brokerage trade or business. The corporation can then use losses from rental estate to offset non passive income such as wages, interest, and dividends.

Net Operating Loss Carrybacks and Carryforwards

If a company has a net operating loss, the loss may be applied against income in other years. The loss can be carried back 3 years and then forward 15 years (however, a capital loss can be carried back for 3 years and forward for only 5 years).

Note that the taxpayer must first apply the loss against the taxable income in the 3 prior years. If the loss is not completely absorbed by the profits in these 3 years, it may be carried forward to each of the following 15 years. The taxpayer has the option of foregoing the carryback and can elect to carry the loss forward only. After 15 years, any remaining loss may no longer be used as a tax deduction.

To illustrate, a 1995 net operating loss may be carried back to the three prior years and used to recover, in whole or in part, the taxes paid during 1992, 1993, and 1994. Thus, the net operating loss can be carried back first to 1992 and treated as a newly discovered deduction to be subtracted from the gross income reported in 1992. If the corporation had taxable income and paid a tax in 1992, the corporation can get a refund of part or all of the taxes paid for that year. If there is still a remaining unabsorbed loss, the corporation can carry it to the second prior year (1993), and then to the first prior year (1994).

Finally, if any part of the loss remains after the carryback to the three prior years, this unabsorbed loss may be used to reduce taxable income, if any, during the 15-year period from 1996 through 2010.

Tax Rates

The federal income tax rate is graduated, meaning that as additional profits are earned, the tax rate on the incremental earnings increases.

The federal corporate tax rates are as follows:

Taxable income	Tax rates
$0—$50,000	15%
$50,001—$75,000	25%
$75,000—$10 million	34%
Over $10 million	35%

There is an additional 5% surtax on taxable income between $100,000 and $335,000. Additionally, the benefit of lower corporate tax brackets is phased out for certain corporations with very high taxable income. A 3% tax is imposed on taxable income over $15 million, up to a maximum of $100,000 in additional tax.

Personal service corporations are not eligible for the graduated rates and must pay tax on their entire taxable income at a flat rate of 35%. A corporation is considered as a personal service corporation if the principal activity of the corporation is the performance of professional services (such as medicine, law, accounting, engineering, etc.) and these services are substantially performed by the employee-owners.

The highest tax rate used in computing a company's total tax obligation is referred to as the marginal tax rate, which is the tax rate applicable to the next dollar of income. In addition to the marginal tax rate, the average (effective) tax rate is the one applicable to all taxable income. It is computed as follows:

$$\text{Effective tax rate} = \frac{\text{Total tax liability}}{\text{Taxable income}}$$

The maximum federal tax rate for corporations is 35%. In addition, companies have state and local income taxes that are typically based on federal taxable income. As a result, the effective tax rate a company pays is generally higher than 35%.

Example 13.7. Your company's taxable income is $160,000. The tax obligation is:

$50,000 × 15%	$ 7,500
$25,000 × 25%	6,250
$85,000 × 34%	28,900
$60,000 × 5%	3,000
	$45,650

The average (effective) tax rate is $45,650/$160,000 = 28.5%.

Because you are concerned with the tax obligation on the additional profit, the marginal tax rate must be considered when making financial decisions.

Tax Prepayments and Credits

Under the principle of pay-as-you-go of our tax system, 90% of the corporation's tax liability must be prepaid. The quarterly estimated payments are to be made on April 15, June 15, September 15, and December 15 for calendar year taxpayers.

A tax credit reduces dollar for dollar the amount of tax otherwise payable. A foreign tax credit is allowed for income taxes paid to a foreign country. However, the foreign tax credit cannot be used to reduce the U.S. tax liability on income from U.S. sources. The allowable credit is calculated as follows:

$$\text{Foreign tax credit} = \frac{\text{Foreign source income}}{\text{Worldwide income}} \times \text{U.S. tax liability}$$

Example 13.8. In 19X8, your company had worldwide taxable income of $675,000 and tentative U.S. income tax of $270,000. The company's taxable income from business operations in Country X was $300,000, and foreign income taxes charged were $135,000 stated in U.S. dollars. The credit for foreign income taxes that can be claimed on the U.S. tax return for 19X8 is:

$$\text{Foreign tax credit} = \frac{\$300,000}{\$675,000} \times \$270,000 = \$120,000$$

TAX STRATEGIES AND PLANNING

You have to analyze the tax consequences of alternative approaches in financial decision making. Ignoring the effect of income taxes will cause an overstatement in estimated income. An otherwise positive cash flow from an investment may become negative when the tax is taken into consideration. As a result, an investment alternative may be chosen that does not sufficiently generate the needed return for the risk exposure taken.

Tax planning includes the minimization of tax payments by attempting to eliminate the tax entirely or providing for taxation at lower tax rates. But reducing the tax ultimately payable is often not possible. As an alternative, much tax planning is concerned with deferring tax payments from the current year to a future tax year. Tax deferrals can contribute a great deal to the efficient use of resources available to the company. The deferral is equivalent to an interest-free loan from the taxing authority.

In addition, the business earns a return for another year on the funds that would have had to be paid to the federal and local taxing authorities; future tax payments will be made with "cheaper" dollars; and there may eventually be no tax payment (e.g. new tax laws). Thus, corporate taxes should be deferred when there will be a lower tax bracket in a future year, or when the firm lacks the funds to meet the present tax obligation.

In tax planning, income and expenses should be shifted into tax years that will result in the least tax liability. In general, income should be deferred to future years and expenses should be accelerated into the current year. This tax strategy is based on the theory that the time value of money makes the deferral of tax liability desirable.

Deferring Taxable income

Deferring or postponing the reporting of taxable income is always advantageous since it enable a corporation to defer taxes and use the funds for an additional period of time. Thus, a good tax strategy is to defer income into a year in which it will be taxed at a lower rate. For example, if you expect tax rates to drop next year, you can reduce the tax obligation by deferring income into the next year. Thus, instead of selling an asset outright, the corporation can exchange that asset for a like-kind asset, and the gain on the exchange will be tax-deferred.

For example, a tangible asset such as equipment, machinery, autos, etc. held for productive use in business or for investment can be exchanged tax-free for another tangible asset. Similarly, realty such as land, buildings, warehouses, etc. can be exchanged tax-free for another piece of realty property.

Another method to defer income is to report the gain from an asset sale under the installment method. This method can be used when the sales price is collected over more than one year. The company must determine a gross profit ratio for an installment sale by dividing the gross profit realized on the sale by the total contract price. This gross profit ratio is then applied to the cash collections each year to determine the amount of gain that is taxable in that year.

Converting Taxable Income into Tax-Exempt Income

The company should try to convert income to less taxed or even tax-exempt sources. For example, a change from investing in corporate bonds into investing in state and local obligations can convert taxable interest income into tax-exempt income. Interest earned on municipal bonds is not subject to federal taxes and is exempt from the tax of the state in which the bond is issued. Of course, the market value of the bond changes with changes in the "going" interest rate.

Tax-free income is worth much more than taxable income. That is why tax-free bonds almost always have lower pre-tax yields. You can determine the equivalent taxable return as follows:

$$\text{Equivalent taxable return} = \frac{\text{Tax-free return}}{1 - \text{Marginal tax rate}}$$

Example 13.9. A municipal bond pays interest at the rate of 6%. The company's marginal tax rate is 34%. The equivalent rate on a taxable instrument is:

$$\frac{.06}{1 - .34} = \frac{.06}{.66} = 9.1\%$$

Accelerating Expenses

Pay tax-deductible expenses in a year in which you will receive the most benefit. Accelerate expenses which will no longer be deductible or will be restricted in the future. Also, accelerate deductions into the current year if you anticipate lower tax rates in the next period. For example, if your company is a cash-basis taxpayer, prepay your future tax obligations, such as property taxes and estimated income taxes, to bring the deductible expenses into the current year. Also, selling a loss asset prior to year-end can accelerate a tax loss into the current year.

The company may donate property instead of cash. By donating appreciated property to a charity, the business can deduct the full market value and avoid paying tax on the gain.

Avoiding Statutory Tax Traps

Corporations may be subject to two penalty taxes: the accumulated earnings tax and the personal holding company tax.

The accumulated earnings tax may be imposed on closely-held corporations (other than S corporations discussed later) that retain profits beyond the "reasonable needs" of the business. The unreasonable portion of accumulated earnings is subject to a penalty tax of 28%. This penalty tax is in addition to any regular or alternative minimum tax liability. Most corporations are allowed a $250,000 credit ($150,000 for personal service corporations) against accumulated taxable income. If it appears that there may be exposure to an accumulated earnings tax penalty, the corporation should consider reducing taxable income by acquiring noninvestment assets, increasing compensation to officer-stockholders, increasing dividends, electing an installment sales method to report income, etc. Also, the corporation should document carefully the reasonable needs of the business.

The personal holding company tax is a penalty tax levied on closely held corporations which receive sizable amounts of investment income such as dividends, interest, and rents. The tax rate is 28% of undistributed personal holding company income. The primary purpose of this penalty tax is to force a company to distribute its earnings to shareholders as dividends if the retained earnings are not invested in operating assets. Companies exposed to the risk of this penalty tax should consider postponing investment income to a future tax year, accelerating operating income, paying increased dividends to officer-shareholders, etc.

Electing Subchapter S Corporation

A regular corporation can elect to become a Subchapter S corporation (more commonly known as an S corporation) in order to be taxed in the same manner as a partnership. In an S corporation, there is no tax at the corporate level and all corporate income and losses are passed through to the stockholders. Income is allocated and taxed directly to the stockholders regardless of whether it is actually received by them. An S corporation files an information tax return on form 1120S, and attaches a Schedule K-1 for each stockholder, showing his/her portion of taxable income or loss for the year. The corporate income is taxed only once to the stockholders at their individual tax rates.

In order to elect subchapter S treatment, a corporation must meet the following qualifications:

Have no more than 35 shareholders.
The shareholders do not include corporations and partnerships.
Have only one class of stock.
All shareholders must be either U.S. citizens or resident aliens.

Each of these criteria must be satisfied before a corporation can make a valid election. Failure to meet any of these conditions during a year will automatically terminate the election.

The election to be taxed as an S corporation may be made by a qualified corporation at any time during the preceding taxable year or in the first 75 days of the year during which it applies. For the election to be valid, all stockholders must consent to the election in a signed statement.

An S corporation is advantageous because there is no tax at the corporate level. Thus, stockholders still obtain the benefits of a corporation such as limited liability while escaping the double taxation typically associated with the distribution of corporate profits. Also, an S shareholder can be an employee of the corporation. As a result, the corporation's income allocated to the shareholders is not subject to self-employment tax. If the corporation has net losses, as is often the case in the early years of newly formed businesses, these losses will also be allocated to the owners who may then deduct them on their individual tax returns. However, certain tax advantages that companies obtain are not given to S corporations.

For example, fringe benefits paid to shareholders owning more than 2% of the stock are treated as distributions to the shareholders and are not deductible by the corporation. In addition, items that would be subject to special limitations on the shareholders' individual returns, such as interest, dividends, capital gains and losses, charitable contributions, etc., cannot be netted against the S corporation's income; instead they must be allocated to the shareholders directly.

CONCLUSION

This chapter attempted to help the engineer and engineering manager obtain a basic understanding of how taxes affect business decisions and to develop an appreciation for tax planning. With a firm grasp of tax fundamentals, the engineer should be better able to identify potential opportunities and trouble areas and effectively incorporate tax factors in decision making. He should also be able to communicate more effectively with the tax specialist.

As an engineer, you must have a basic understanding of taxation as well as accounting and finance. The financial functions of business involve record keeping, performance evaluation, variance analysis, budgeting, and utilization of resources. To optimally perform your duties, you must also have a good understanding of taxation as related to accounting and finance. Without a firm grasp of all these disciplines, you do not have the tools needed for effective financial decision making. Decisions which may make sense in marketing and sales must also make financial sense. You must have the background to give sound input into the decision process.

The engineer and project manager should try to obtain the most tax benefit for the company. The tax effect associated with new projects and specific financial and investment decisions has to be considered. Through effective tax planning, the corporation can save much needed working capital to survive the economic uncertainties of the 1990s.

It is important, however, to realize that tax laws are complex and change frequently. Therefore, you are strongly advised to review the potential effect of taxation with the company accountant and tax professional before deciding on any particular course of action.

Figure 13.1 provides a brief summary of the new tax code.

Higher tax rates:
- Individuals from 31% to 36%, with possible effective rate of 42%.
- Corporations from 34% to 35%.
- Capital gains stays at 28%.

Investment incentive:
- Only half of gain on stock taxed if held for more than five years. Applies only to investments by noncorporate taxpayers who purchase originally issued stock in companies with $50 million or less in assets, prior to subtracting short-term indebtedness.
- Gain eligible for exclusion is limited to $10 million or 10 times stock's value for tax purposes, whichever is greater.

Executives targeted:
- Retirement and savings plans can cater to executives only up to $150,000 of their salary before endangering plan's tax-qualified status.
- Deny deduction for executive's compensation in excess of $1 million at publicly held companies. *Exceptions* include "commissions" and performance-related compensation "if certain outside director and shareholder approval requirements are met."

Other deductions disallowed:
- Dues for clubs, including airline and hotel clubs.
- Cost of lobbying federal or state legislation or cost of attempting to influence "the official actions or position" of federal officials. Does not apply to lobbying of "local council or similar governing body."

Deductions reduced:
- Business meals and entertainment from 80% to 50%.
- That portion of dues paid to tax-exempt organizations that is used for lobbying or other political activity.

Intangibles codified:
- Fifteen-year (reduced from 25 years) amortization of intangibles whether or not acquired as part of the acquisition. Intangibles include "good-will" and "going concern" value; work force, information base, know-how, customer and supplier-based intangibles; licenses, permits and other rights granted by governmental units; covenants not to compete, and franchises, trademarks and trade names.
- Effective as of enactment, but may be retroactively applied to intangibles acquired after July 25, 1991.

Figure 13.1 Decoding the new tax code. (*Source*: Dow, Lohnes & Albertson.)

GLOSSARY

Accelerated depreciation A depreciation method wherein the depreciation charges decrease with time.

Accounting equation An expression of the equivalency in dollar amounts of assets and liabilities and equity in *double-entry accounting*, often stated as Assets = Liabilities + owners' equity.

Accounting profit Net income or earnings figures shown on the income statement.

Accounts receivable turnover Annual credit sales divided by average accounts receivable.

Accrual basis accounting Recognition of revenue when earned and expenses when incurred.

After-tax cash flow The net cash flow (cash revenue less cash expenses) after taxes have been subtracted. It is the cash flow generated from operations.

Amortization The spreading out of costs over a period of time.

Amortized loan A loan that is paid off in periodic equal installments and includes varying portions of principal and interest during its term.

Analysis of variances (variance analysis) An analysis and investigation of causes for variances between standard costs and actual costs. A variance is considered favorable if actual costs are less than standard costs; it is unfavorable if actual costs exceed standard costs. Unfavorable variances are the ones that need further investigation into their causes, so corrective action may be taken.

Annual report A glossary magazine-style report that companies must send to shareholders annually. It includes the company's financial statements.

Annuity A series of equal periodic payments or receipts.

Annuity due An annuity where payments or receipts occur at the beginning of the period.

Asset turnover The sales divided by average total assets, revealing the efficiency of assets in generating revenue.

Assets The resources owned by the firm.

Balance sheet A table that shows the status of a company's assets, liabilities, and owners' equity as of a given date.

Balance sheet equation See *Accounting equation.*

Bond A form of interest-bearing note payable, or simply a corporate long term I.O.U.

Book value per share The worth of each share of stock per the books based on historical cost.

Break-even analysis A branch of cost-volume-profit (CVP) analysis that determines the break-even sales, which is the level of sales where total costs equal total revenue. At the break-even point, there is no profit or loss.

Break-even point The level of sales where total costs equal total revenue.

Budget A quantitative plan of activities and programs expressed in terms of assets, liabilities, revenues, and expenses. See also *Master budget.*

Capital budget A budget or plan of proposed acquisitions and replacements of long-term assets and their financing. A capital budget is developed using a variety of capital budgeting techniques such as the discount cash flow method.

Capital budgeting The process of making long-term planning decisions for capital investments.

Capital rationing The selection of the mix of acceptable projects that provides the highest overall net present value of future cash flows when a company has a limit on the budget for capital spending.

Capital structure Composition of common stock, preferred stock, retained earnings, and long-term debt maintained by the business entity in financing its assets.

Cash basis accounting Method of recognizing revenue and expenses when cash is received or paid.

Cash budget A budget for cash planning and control that presents anticipated cash inflow and cash outflow for a specified time period. The cash budget helps the owner keep cash balances in reasonable relationship to needs. It assists in avoiding idle cash and possible cash shortages. The cash budget shows beginning cash, cash receipts, cash payments, and ending cash.

Cash flow 1. Cash receipts minus cash disbursements from a given operation or asset for a given period. Cash flow and cash inflow are often used interchangeably. 2. The monetary value of the expected benefits and costs of a project. It may be in the form of cash savings in operating costs or the difference between additional dollars received and additional dollars paid out for a given period.

Cash flow statement A statement showing from what sources cash has come into the business and on what the cash has been spent. Cash flow is broken down into operating, investing, and financing activities.

Chart of accounts A list of account names and numbers found in the general ledger and arranged in the order in which they customarily appear in the financial statements.

Chief Financial Officer (CFO) Executive who directs all financial aspects of the business.

Collection period The number of days it takes to collect accounts receivable. It equals 365 days divided by the accounts receivable turnover. The collection period should be compared to the terms of sale.

Common costs A cost shared by different departments, products, or jobs. Also called *joint costs* or *indirect costs*.

Comprehensive budget See *Master budget*.

Constraining (limiting) factor The item or factor that restricts or limits production or sale of a given product. Virtually all firms suffer from one or more constraining factors. Examples of constraining factors include limited machine-hours, labor-hours and shortage of materials and skilled labor. Other limiting factors may be cubic feet of display or warehouse space, or working capital.

Contribution approach to pricing An approach to pricing a special order. This situation occurs because a company often receives a nonroutine, special order for its products at lower prices than usual. In normal times, the company may refuse such an order since it will not yield a satisfactory profit. If times are bad or when there is idle capacity, an order should be accepted if the incremental revenue exceeds the incremental costs involved.

Contribution margin (CM) The difference between sales and the variable costs of the product or service, also called marginal income. It is the amount of money available to cover fixed costs and generate profits.

Contribution (margin) income statement An income statement that organizes the costs by behavior. It shows the relationship of variable costs and fixed costs, regardless of the functions that a given cost item is associated with.

Contribution margin (CM) ratio The contribution margin (cm) as a percentage of sales.

Controller Chief accounting executive of an organization. The controller is in charge of the Accounting Department. The principal functions of the controller are: (1) planning for control, (2) financial reporting and interpreting, (3) tax administration, (4) management audits and development of accounting systems, and (5) internal audits.

Conversion costs The sum of the costs of direct labor and factory overhead.

Cost accounting A system for recording and reporting measurements of the cost of manufacturing goods and performing services in the aggregate and in detail. It includes methods for reorganizing, classifying, allocating, aggre-gating and reporting actual costs and comparing them with standard costs.

Cost accumulation The collection of costs in an organized fashion by means of a cost accounting system. There are two primary approaches to cost accu-mulation: a job order system and a process cost system.

Cost behavior analysis An analysis of mixed costs. Mixed costs must be sepa-rated into variable and fixed elements in order to be included in a variety of business planning analyses such as cost-volume-profit (CVP) analysis.

Cost of capital The minimum rate of return that is necessary to keep the market value (or stock price) of a firm.

Cost center The unit within the organization in which the manager is responsi-ble only for costs. A cost center has no control over sales or over the gener-ating of revenue. An example is the production department of a manufac-turing company. The performance of the cost center is evaluated by comparing actual costs to budgeted costs.

Cost control The steps taken by management to assure that the cost objectives set down in the planning stage are attained, and to assure that all segments of the organization function in a manner consistent with its policies. For effective cost control, most organizations use standard cost systems, in which the actual costs are compared against standard costs for performance evaluation and the deviations are investigated for remedial actions. Cost control is also concerned with feedback that might change any or all of the future plans, the production method, or both.

Cost of production report A summary of the unit and cost data of a production department in a process cost system.

Cost-volume formula A cost function in the form of $y = a + bx$. For example, the cost-volume formula for factory overhead is $y = \$200 + \$10x$ where $y =$ estimated factory overhead and $x =$ direct labor hours, which means that the factory overhead is estimated to be $200 fixed, plus $10 per hour of direct labor. Cost analysts use the formula for cost prediction and flexible budget-ing purposes.

Cost-volume-profit (CVP) analysis An analysis that deals with how profits and costs change with a change in volume. It looks at the effects on profits of changes in such factors as variables costs, fixed costs, selling prices, volume, and mix of products sold.

Credit entry Transaction recorded on the right hand side of an account.

Current ratio Current assets divided by current liabilities.

Debit entry Transaction recorded on the left hand side of an account.

Debt-equity ratio Total liabilities divided by total stockholders' equity.

Departmental rate A predetermined factory overhead rate for each production department.

Depreciation The procedure of spreading out the acquisition cost of fixed assets (such as machinery and equipment) to each of the time periods in which they are utilized.

Direct labor budget A schedule for expected labor cost. Expected labor cost is dependent upon expected production volume (production budget). Labor requirements are based on production volume multiplied by direct labor hours per unit. Direct labor hours needed for production is then multiplied by direct labor cost per hour to derive budgeted direct labor costs.

Direct materials budget A budget that shows how much material will be required for production and how much material must be bought to meet this production requirement. The purchase depends on both expected usage of materials and inventory levels.

Discounted cash flow (DCF) techniques Methods of selecting and ranking investment proposals such as the net present value (NPV) and internal rate of return (IRR) methods where the time value of money is taken into account.

Discretionary (fixed) costs The fixed costs that change because of managerial decisions, also called *management (fixed) costs* or *programmed (fixed) costs*. Examples of this type of fixed cost are advertising, training, and research and development.

Double entry bookkeeping A method of accounting that recognizes the duality of a transaction such that any a change in one account also causes a change in another account.

Dupont formula The breakdown of return on investment (ROI) into profit margin and asset turnover.

Earnings A synonym for net income or profit after taxes.

Earnings per share (EPS) Net income after taxes available to common shareholders on a per share basis.

Economic profit The same as cash inflows; see *Cash flow*.

Equity The difference between the assets of an entity and its liabilities. Also called *net worth, owners' equity*, or *stockholders' equity*.

Facilities investment budget A budget plan prepared for individual facility expenditure projects. The time span of this budget depends upon the project. Capital expenditures to be budgeted include replacement, acquisition, or construction of plants and major equipment. See also *Capital budgeting*.

Factory overhead budget A schedule of all expected manufacturing costs except for direct materials and direct labor. Factory overhead items include indirect materials, indirect labor, factory rent, and factory insurance.

Financial Accounting Standards Board (FASB) The seven-member board which currently has the authority to formulate and issue pronouncements of *generally accepted accounting principles*.

Financial leverage A portion of a firm's assets financed with debt instead of equity.

Fixed budget See *Static budget*.

Fixed cost A cost that remains the same for each period in the short run regardless of activity. Examples are rent, interest, insurance, and property taxes.

Flexible budget A budget based on cost-volume relationships and developed for the actual level of activity. It is an extremely useful tool for comparing the actual cost incurred to the cost allowable for the activity level achieved.

Flexible budget formula See *Cost-volume formula*.

Fundamental equation of accounting See *Accounting equation*.

General ledger A grouping of the accounts in which the transactions of an entity are recorded.

Generally accepted accounting principles (GAAP) Rules, regulations, standards, conventions, and pronouncements used as a basis for financial reporting, and in the preparation of financial statements.

Gross profit margin The ratio of gross profit to net sales. A high gross profit margin is a positive sign since it shows the business is earning an attractive return over the cost of its merchandise sold.

Illiquid 1. Lacking enough liquid assets, like cash and marketable securities, to cover short-term obligations. 2. Current liabilities exceed current assets.

Income statement A table that details a company's revenue, expenses, and profit or loss for a given period.

Industry norm A typical ratio for the industry based on averaging companies' values.

Insolvency The failure of a company to meet its obligations as they become due. An analysis of insolvency concentrates on the operating and capital structure of the business. The proportion of long-term debt in the capital structure must also be considered.

Internal rate of return (IRR) The rate of return on a proposal that equates the initial investment with the present value of future cash inflows.

Inventory turnover The number of times inventory is sold during the year. It equals cost of goods sold divided by the average dollar balance. Average inventory equals the beginning and ending balances divided by two.

Investment center A responsibility center within an organization that has control over revenue, cost and investment funds. It is a profit center whose performance is evaluated on the basis of the return earned on invested capital.

Job order costing The accumulation of costs by specific jobs, contracts, or orders. This costing method is appropriate when direct costs can be identified with specific units of production. Job order costing is widely used by custom manufacturers such as printing, aircraft, construction, auto repair and professional services.

Joint costs All the common manufacturing costs incurred prior to the point, referred to as the split-off point, where the joint products are identified as individual products.

Joint products The products that have a relatively significant sales value when two or more types of products are produced simultaneously from the same input by a joint process. For example, gasoline, fuel oil, kerosene, and paraffin are the joint products that are produced from crude oil.

Journal A book in which all business transactions are recorded in chronological order.

Just-in-time (JIT) A demand-pull system where demand for customer output (not plans for using input resources) triggers production. Production activities are "pulled", not "pushed," into action.

Labor efficiency variance The difference between the amount of labor time that should have been used and the labor that were actually used, multiplied by the standard rate.

Labor rate variance A deviation from standard in the average hourly rate paid to workers.

Least-squares method The method that fits a line in such a way that the sum of squared distances from the data points to the line is minimized.

Ledger Book in which all accounts of the business are maintained.

Liabilities Obligations of the firm to outside creditors such as lenders and bondholders.

Liquidity The ability of current assets to meet current liabilities when due.

Management by exception A management concept or policy by which management devotes its time to investigating only those situations in which actual results differ significantly from planned results. The idea is that management should spend its valuable time concentrating on the more important items (such as the shaping of the company's future strategic course).

Master (comprehensive) budget A plan of activities expressed in monetary terms of the assets, equities, revenues, and costs which will be involved in carrying out the plans. It is a set of projected or planned financial statements.

Materials price variance The difference between what is paid for a given quantity of materials and what should have been paid, multiplied by the actual quantity of materials used.

Materials quantity (usage) variance The difference between the actual quantity of materials used in production and the standard quantity of materials allowed for actual production, multiplied by the standard price per unit.

Mixed costs The costs that vary with changes in volume but, unlike variable costs, do not vary in direct proportion, also called *semivariable costs.*

Modified Accelerated Cost Recovery System (MACRS) The system used in computing annual depreciation for assets acquired in 1987.

Net present value (NPV) The difference between the present value of cash inflows generated by the project and the amount of the initial investment.

Notes to the financial statements Important information supplementing the financial statements.

Operating cycle The average time period between buying inventory and receiving cash proceeds from its eventual sale. It is determined by adding the number of days inventory is held and the collection period for accounts receivable.

Operating leverage The degree to which the firm chooses to lock in fixed costs other than financing costs.

Operational (operating) budget A budget that embraces the impacts of operating decisions. It contains forecasts of sales, net income, the cost of goods sold, selling and administrative expenses, and other expenses.

Opportunity cost The revenue forfeited by rejecting an alternative use of time or facilities.

Ordinary annuity An annuity where payments or receipts occur at the end of the period.

Out-of-pocket cost The actual cash outlays made during the period for payroll, advertising, and other operating expenses. Depreciation is not an out-of-pocket cost, since it involves no current cash expenditure.

Owners' equity The value of the firm to its owners. It is the difference between assets and liabilities. See also *Equity*.

Payback period The number of years it takes to recover your initial investment. The payback period equals the initial investment divided by the annual cash inflow.

Planning The selection of short- and long-term objectives and the drawing up of tactical and strategic plans to achieve those objectives. In planning, managers outline the steps to be taken in moving the organization toward its objectives. After deciding on a set of strategies to be followed, the organization needs more specific plans, such as locations, methods of financing, hours of operation, etc. As these plans are made, they will be communicated throughout the organization. When implemented, the plans will serve to coordinate, or meld together, the efforts of all parts of the organization toward the company's objectives.

Predetermined overhead rates An overhead rate, based on budgeted factory overhead cost and budgeted activity, that is established before a period begins.

Preferred stock An equity instrument that promises to pay its holders a fixed dividend.

Present value The current worth of future sums of money.

Pro forma balance sheet A budgeted balance sheet.

Pro forma income statement A budgeted income statement.

Process costing A cost accumulation method used to assign costs to units of a homogeneous product as the units pass through one of more processes.

Production budget A schedule for expected units to be produced. It sets forth the units expected to be manufactured to satisfy budgeted sales and inventory requirements. Expected production volume is determined by adding desired ending inventory to planned sales and then subtracting beginning inventory.

Product mix See *Sales mix.*

Profitability index The ratio of the total present value of future cash inflows to the initial investment.

Profit margin The ratio of net income to net sales. It reveals the entity's ability to generate profit at a given sales level. The ratio gives the owner an indicator of the operating efficiency and pricing strategy of the business.

Profit planning A process of developing a profit plan which outlines the planned sales revenues and expenses and the net income or loss for a time period. Profit planning requires preparation of a master budget and various analyses for risk and "what-if" scenarios. Tools for profit planning include the cost-volume-profit (CVP) analysis and budgeting.

Profit-volume chart A chart that determines how profits vary with changes in volume.

Projected (budgeted) balance sheet A schedule for expected assets, liabilities, and stockholders' equity. It projects a company's financial position as of the end of the budgeting year. The reasons for preparing a budgeted balance sheet are that it: (1) discloses unfavorable financial condition that management may want to avoid, (2) serves as a final check on the mathematical accuracy of all other budgets, and (3) highlights future resources and obligations.

Projected (budgeted) income statement A summary of various component projections of revenues and expenses for the budget period. It indicates the expected net income for the period.

Quick ratio Also called *acid-test ratio,* the most liquid current assets (cash, marketable securities, accounts receivable) divided by current liabilities.

Rate of return on investment (ROI) 1. For the company as a whole, net income after taxes divided by invested capital. 2. For a segment of an organization, net operating income divided by operating assets. 3. For capital budgeting purposes, also called *simple, accounting,* or *unadjusted rate of return,* expected future net income divided by initial (or average) investment.

Regression analysis A statistical procedure for estimating mathematically the average relationship between the dependent variable (sales, for example) and one or more independent variables (price and advertising, for example).

Relevant cost The expected future cost that will differ between the alternatives being considered.

Residual income (RI) The operating income which an investment center is able to earn above some minimum return on its assets. It equals operating income less the minimum rate of return times total assets.

Responsibility accounting The collection, summarization, and reporting of financial information about various decision centers (responsibility centers) throughout an organization; also called activity accounting or profitability accounting.

Responsibility center A unit in the organization which has control over costs, revenues, or investment funds. For accounting purposes, responsibility centers are classified as cost centers, revenue centers, profit centers, and investment centers, depending on what each center is responsible for.

Retained earnings The amount of income realized and retained since the inception of the entity less dividends paid out to shareholders.

Return The reward for making an investment in the form of earnings and appreciation in value.

Risk 1. Variability about income, returns, or other financial variable. 2. The possibility of losing value.

Risk-return trade-off A comparison of the expected return from an investment with the risk associated with it. The higher the risk undertaken, the more ample the return. Conversely, the lower the risk, the more modest the return.

Rule of OPM The use of other people's money (OPM) to magnify potential returns from the business. It is hoped that investment through leverage will earn a rate of return greater than the costs of borrowing. Also called *trading on the equity*.

Sales mix The relative proportions of the product sold.

Sales budget An operating plan for a period expressed in terms of sales volume and selling prices for each class of product or service. Preparation of a sales budget is the starting point in budgeting since sales volume influences nearly all other items.

Sales forecasting A projection or prediction of future sales. It is the foundation for the quantification of the entire business plan and a master budget. Sales forecasts serve as a basis for planning.

Scatter diagram A plot of data points in a X-Y graph.

Segmented reporting The process of reporting activities of various segments of an organization such as divisions, departments, product lines, services, sales territories, or customers.

Simple regression A regression analysis which involves one independent variable. For example, total factory overhead is related to one activity variable (either direct labor hours or machine hours).

Spreadsheet A working paper having numbers in rows and columns.

Standard A quantitative expression of a performance objective, such as standard hours of labor allowed for actual production or a standard purchase price of materials per unit. Sometimes the terms standard and budget are used interchangeably.

Standard costs Production or operating costs that are carefully predetermined. A standard cost is a target cost that should be attained.

Standard cost system A system by which production activities are recorded at standard costs and variances from actual costs are isolated.

Standard hours allowed The standard time that should have been used to manufacture actual units of output during a period. It is obtained by multiplying actual units of production by the standard labor time.

Standard labor rate The standard rate for direct labor that would include not only base wages earned but also an allowance for fringe benefits and other labor-related costs.

Standard materials price The standard price per unit for direct materials. It reflects the final, delivered cost of the materials, after any discounts are taken.

Standard quantity allowed The standard amount of materials that should have been used to manufacture units of output during a period. It is obtained by multiplying actual units of production by the standard material quantity per unit.

Static (fixed) budget A budget based on the anticipated output level rather than on the actual attained output level.

Strategic planning The implementation of an organization's objectives. Strategic planning decisions will have long-term impacts on the organization while operational decisions are day-to-day in nature.

Time value of money The value of money at different time periods. As a rule, one dollar today is worth more than one dollar tomorrow. The time value of money is a critical consideration in financial decisions.

Treasurer Financial officer in a firm who deals with money problems.

Turnover The number of times an asset, such as inventory, turns over during an accounting period.

Uniform annual cost (UAC) The uniform end of period cost which is equivalent to a present sum. For UAC calculations, cash flows such as an initial investment, salvage value, and annual operating cost are spread out over the life of the project by converting them into an equivalent uniform annual series.

Variance In cost analysis, the deviation between the actual cost and the standard cost.

Zero-base budgeting A method of budgeting in which cost and benefit estimates are built up from scratch, from the zero level, and must be justified.

APPENDIX

INTEREST TABLES

Table A1 The Future Value of $1.00 (Compounded Amount of $1.00)

$(1 + i)^n = T1(i,n)$

Periods	4%	6%	8%	10%	12%	14%	20%
1	1.040	1.060	1.080	1.100	1.120	1.140	1.200
2	1.082	1.124	1.166	1.210	1.254	1.300	1.440
3	1.125	1.191	1.260	1.331	1.405	1.482	1.728
4	1.170	1.263	1.361	1.464	1.574	1.689	2.074
5	1.217	1.338	1.469	1.611	1.762	1.925	2.488
6	1.265	1.419	1.587	1.772	1.974	2.195	2.986
7	1.316	1.504	1.714	1.949	2.211	2.502	3.583
8	1.369	1.594	1.851	2.144	2.476	2.853	4.300
9	1.423	1.690	1.999	2.359	2.773	3.252	5.160
10	1.480	1.791	2.159	2.594	3.106	3.707	6.192
11	1.540	1.898	2.332	2.853	3.479	4.226	7.430
12	1.601	2.012	2.518	3.139	3.896	4.818	8.916
13	1.665	2.133	2.720	3.452	4.364	5.492	10.699
14	1.732	2.261	2.937	3.798	4.887	6.261	12.839
15	1.801	2.397	3.172	4.177	5.474	7.138	15.407
16	1.873	2.540	3.426	4.595	6.130	8.137	18.488
17	1.948	2.693	3.700	5.055	6.866	9.277	22.186
18	2.026	2.854	3.996	5.560	7.690	10.575	26.623
19	2.107	3.026	4.316	6.116	8.613	12.056	31.948
20	2.191	3.207	4.661	6.728	9.646	13.743	38.338
30	3.243	5.744	10.063	17.450	29.960	50.950	237.380
40	4.801	10.286	21.725	45.260	93.051	188.880	1469.800

Table A2 The Future Value of an Annuity of $1.00* (Compounded Amount of an Annuity of $1.00)

$$\frac{(1 + i)^n - 1}{i} = T2(i,n)$$

Periods	4%	6%	8%	10%	12%	14%	20%
1	1.000	1.000	1.000	1.000	1.000	1.000	1.000
2	2.040	2.060	2.080	2.100	2.120	2.140	2.200
3	3.122	3.184	3.246	3.310	3.374	3.440	3.640
4	4.247	4.375	4.506	4.641	4.779	4.921	5.368
5	5.416	5.637	5.867	6.105	6.353	6.610	7.442
6	6.633	6.975	7.336	7.716	8.115	8.536	9.930
7	7.898	8.394	8.923	9.487	10.089	10.730	12.916
8	9.214	9.898	10.637	11.436	12.300	13.233	16.499
9	10.583	11.491	12.488	13.580	14.776	16.085	20.799
10	12.006	13.181	14.487	15.938	17.549	19.337	25.959
11	13.486	14.972	16.646	18.531	20.655	23.045	32.150
12	15.026	16.870	18.977	21.385	24.133	27.271	39.580
13	16.627	18.882	21.495	24.523	28.029	32.089	48.497
14	18.292	21.015	24.215	27.976	32.393	37.581	59.196
15	20.024	23.276	27.152	31.773	37.280	43.842	72.035
16	21.825	25.673	30.324	35.950	42.753	50.980	87.442
17	23.698	28.213	33.750	40.546	48.884	59.118	105.930
18	25.645	30.906	37.450	45.600	55.750	68.394	128.120
19	27.671	33.760	41.446	51.160	63.440	78.969	154.740
20	29.778	36.778	45.762	57.276	75.052	91.025	186.690
30	56.085	79.058	113.283	164.496	241.330	356.790	1181.900
40	95.026	154.762	259.057	442.597	767.090	1342.000	7343.900

ªPayments (or receipts) at the *end* of each period.

Table A3 Present Value of \$1.00 [T3(i,n)]

Periods	4%	5%	6%	8%	10%	12%	14%	16%	18%	20%	22%	24%	26%	28%	30%	40%
1	0.962	0.952	0.943	0.926	0.909	0.893	0.877	0.862	0.847	0.833	0.820	0.806	0.794	0.781	0.769	0.714
2	0.925	0.907	0.890	0.857	0.826	0.797	0.769	0.743	0.718	0.694	0.672	0.650	0.630	0.610	0.592	0.510
3	0.889	0.864	0.840	0.794	0.751	0.712	0.675	0.641	0.609	0.579	0.551	0.524	0.500	0.477	0.455	0.364
4	0.855	0.823	0.792	0.735	0.683	0.636	0.592	0.552	0.516	0.482	0.451	0.423	0.397	0.373	0.350	0.260
5	0.822	0.784	0.747	0.681	0.621	0.567	0.519	0.476	0.437	0.402	0.370	0.341	0.315	0.291	0.269	0.186
6	0.790	0.746	0.705	0.630	0.564	0.507	0.456	0.410	0.370	0.335	0.303	0.275	0.250	0.227	0.207	0.133
7	0.760	0.711	0.665	0.583	0.513	0.452	0.400	0.354	0.314	0.279	0.249	0.222	0.198	0.178	0.159	0.095
8	0.731	0.677	0.627	0.540	0.467	0.404	0.351	0.305	0.266	0.233	0.204	0.179	0.157	0.139	0.123	0.068
9	0.703	0.645	0.592	0.500	0.424	0.361	0.308	0.263	0.225	0.194	0.167	0.144	0.125	0.108	0.094	0.048
10	0.676	0.614	0.558	0.463	0.386	0.322	0.270	0.227	0.191	0.162	0.137	0.116	0.099	0.085	0.073	0.035
11	0.650	0.585	0.527	0.429	0.350	0.287	0.237	0.195	0.162	0.135	0.112	0.094	0.079	0.066	0.056	0.025
12	0.625	0.557	0.497	0.397	0.319	0.257	0.208	0.168	0.137	0.112	0.092	0.076	0.062	0.052	0.043	0.018
13	0.601	0.530	0.469	0.368	0.290	0.229	0.182	0.145	0.116	0.093	0.075	0.061	0.050	0.040	0.033	0.013
14	0.577	0.505	0.442	0.340	0.263	0.205	0.160	0.125	0.099	0.078	0.062	0.049	0.039	0.032	0.025	0.009
15	0.555	0.481	0.417	0.315	0.239	0.183	0.140	0.108	0.084	0.065	0.051	0.040	0.031	0.025	0.020	0.006
16	0.534	0.458	0.394	0.292	0.218	0.163	0.123	0.093	0.071	0.054	0.042	0.032	0.025	0.019	0.015	0.005
17	0.513	0.436	0.371	0.270	0.198	0.146	0.108	0.080	0.060	0.045	0.034	0.026	0.020	0.015	0.012	0.003
18	0.494	0.416	0.350	0.250	0.180	0.130	0.095	0.069	0.051	0.038	0.028	0.021	0.016	0.012	0.009	0.002
19	0.475	0.396	0.331	0.232	0.164	0.116	0.083	0.060	0.043	0.031	0.023	0.017	0.012	0.009	0.007	0.002
20	0.456	0.377	0.312	0.215	0.149	0.104	0.073	0.051	0.037	0.026	0.019	0.014	0.010	0.007	0.005	0.001
21	0.439	0.359	0.294	0.199	0.135	0.093	0.064	0.044	0.031	0.022	0.015	0.011	0.008	0.006	0.004	0.001
22	0.422	0.342	0.278	0.184	0.123	0.083	0.056	0.038	0.026	0.018	0.013	0.009	0.006	0.004	0.003	0.001
23	0.406	0.326	0.262	0.170	0.112	0.074	0.049	0.033	0.022	0.015	0.010	0.007	0.005	0.003	0.002	
24	0.390	0.310	0.247	0.158	0.102	0.066	0.043	0.028	0.019	0.013	0.008	0.006	0.004	0.003	0.002	
25	0.375	0.295	0.233	0.146	0.092	0.059	0.038	0.024	0.016	0.010	0.007	0.005	0.003	0.002	0.001	
26	0.361	0.281	0.220	0.135	0.084	0.053	0.033	0.021	0.014	0.009	0.006	0.004	0.002	0.002	0.001	
27	0.347	0.268	0.207	0.125	0.076	0.047	0.029	0.018	0.011	0.007	0.005	0.003	0.002	0.001	0.001	
28	0.333	0.255	0.196	0.116	0.069	0.042	0.026	0.016	0.010	0.006	0.004	0.002	0.002	0.001	0.001	
29	0.321	0.243	0.185	0.107	0.063	0.037	0.022	0.014	0.008	0.005	0.003	0.002	0.001	0.001	0.001	
30	0.308	0.231	0.174	0.099	0.057	0.033	0.020	0.012	0.007	0.004	0.003	0.002	0.001	0.001	0.001	
40	0.208	0.142	0.097	0.046	0.022	0.011	0.005	0.003	0.001	0.001						

Table A4 Present Value of Annuity of $1.00 [T4(*i*,*n*)]

Periods	4%	5%	6%	8%	10%	12%	14%	16%	18%	20%	22%	24%	26%	28%	30%	40%
1	0.962	0.952	0.943	0.926	0.909	0.893	0.877	0.862	0.847	0.833	0.820	0.806	0.794	0.781	0.769	0.714
2	1.886	1.859	1.833	1.783	1.736	1.690	1.647	1.605	1.566	1.528	1.492	1.457	1.424	1.392	1.361	1.224
3	2.775	2.723	2.673	2.577	2.487	2.402	2.322	2.246	2.174	2.106	2.042	1.981	1.923	1.868	1.816	1.589
4	3.630	3.546	3.465	3.312	3.170	3.037	2.914	2.798	2.690	2.589	2.494	2.404	2.320	2.241	2.166	1.879
5	4.452	4.330	4.212	3.993	3.791	3.605	3.433	3.274	3.127	2.991	2.864	2.745	2.635	2.532	2.436	2.035
6	5.242	5.076	4.917	4.623	4.355	4.111	3.889	3.685	3.498	3.326	3.167	3.020	2.885	2.759	2.643	2.168
7	6.002	5.786	5.582	5.206	4.868	4.564	4.288	4.039	3.812	3.605	3.416	3.242	3.083	2.937	2.802	2.263
8	6.733	6.463	6.210	5.747	5.335	4.968	4.639	4.344	4.078	3.837	3.619	3.421	3.241	3.076	2.925	2.331
9	7.435	7.108	6.802	6.247	5.759	5.328	4.946	4.607	4.303	4.031	3.786	3.566	3.366	3.184	3.019	2.379
10	8.111	7.722	7.360	6.710	6.145	5.650	5.216	4.833	4.494	4.192	3.923	3.682	3.465	3.269	3.092	2.414
11	8.760	8.306	7.887	7.139	6.495	5.988	5.453	5.029	4.656	4.327	4.035	3.776	3.544	3.335	3.147	2.438
12	9.385	8.863	8.384	7.536	6.814	6.194	5.660	5.197	4.793	4.439	4.127	3.851	3.606	3.387	3.190	2.456
13	9.986	9.394	8.853	7.904	7.103	6.424	5.842	5.342	4.910	4.533	4.203	3.912	3.656	3.427	3.223	2.468
14	10.563	9.899	9.295	8.244	7.367	6.628	6.002	5.468	5.008	4.611	4.265	3.962	3.695	3.459	3.249	2.477
15	11.118	10.380	9.712	8.559	7.606	6.811	6.142	5.575	5.092	4.675	4.315	4.001	3.726	3.483	3.268	2.484
16	11.652	10.838	10.106	8.851	7.824	6.974	6.265	5.669	5.162	4.730	4.357	4.033	3.751	3.503	3.283	2.489
17	12.166	11.274	10.477	9.122	8.022	7.120	6.373	5.749	5.222	4.775	4.391	4.059	3.771	3.518	3.295	2.492
18	12.659	11.690	10.828	9.372	8.201	7.250	6.467	5.818	5.273	4.812	4.419	4.080	3.786	3.529	3.304	2.494
19	13.134	12.085	11.158	9.604	8.365	7.366	6.550	5.877	5.316	4.844	4.442	4.097	3.799	3.539	3.311	2.496
20	13.590	12.462	11.470	9.818	8.514	7.469	6.623	5.929	5.353	4.870	4.460	4.110	3.808	3.546	3.316	2.497
21	14.029	12.821	11.764	10.017	8.649	7.562	6.687	5.973	5.384	4.891	4.476	4.121	3.816	3.551	3.320	2.498
22	14.461	13.163	12.042	10.201	8.772	7.645	6.743	6.011	5.410	4.909	4.488	4.130	3.822	3.556	3.323	2.498
23	14.857	13.489	12.303	10.371	8.883	7.718	6.792	6.044	5.432	4.925	4.499	4.137	3.827	3.559	3.325	2.499
24	15.247	13.799	12.550	10.529	8.985	7.784	6.835	6.073	5.451	4.937	4.507	4.143	3.831	3.562	3.327	2.499
25	15.622	14.094	12.783	10.675	9.077	7.843	6.873	6.097	5.467	4.948	4.514	4.147	3.834	3.564	3.329	2.499
26	15.983	14.375	13.003	10.810	9.161	7.896	6.906	6.118	5.480	4.956	4.520	4.151	3.837	3.566	3.330	2.500
27	16.330	14.643	13.211	10.935	9.237	7.943	6.935	6.136	5.492	4.964	4.525	4.154	3.839	3.567	3.331	2.500
28	16.663	14.898	13.406	11.051	9.307	7.984	6.961	6.152	5.502	4.970	4.528	4.157	3.840	3.568	3.331	2.500
29	16.984	15.141	13.591	11.158	9.370	8.022	6.983	6.166	5.510	4.975	4.531	4.159	3.841	3.569	3.332	2.500
30	17.292	15.373	13.765	11.258	9.427	8.055	7.003	6.177	5.517	4.979	4.534	4.160	3.842	3.569	3.332	2.500
40	19.793	17.159	15.046	11.925	9.779	8.244	7.105	6.234	5.548	4.997	4.544	4.166	3.846	3.571	3.333	2.500

INDEX

About the Authors

JAE K. SHIM is Professor of Accounting and Finance at California State University, Long Beach. An accounting and financial consultant, Dr. Shim is the author or coauthor of over 35 college texts and professional books and numerous referred articles published in journals such as *Financial Management, Management Accounting,* and *Advances in Accounting.* He is a recipient of the 1982 Credit Research Foundation's Outstanding Paper Award for his article on cash budgeting. Dr. Shim received his M.B.A. degree (1968) in business economics/accounting and the Ph.D. degree (1973) in finance/accounting from the University of California, Berkeley.

NORMAN HENTELEFF is President of the National Business Review Foundation, Long Beach, California, and a consultant in engineering management. Dr. Henteleff has taught at several universities and colleges and served as Academic Dean at South Bay College, Los Angeles, California. The coauthor of two books, he is a member of Sigma Iota Epsilon, Phi Delta Kappa, and Delta Pi Epsilon. He received the B.S. (1971) and M.A. (1973) degrees in engineering management from Arizona State University, Tempe, and the Ph.D. degree (1976) in education from the University of Southern California, Los Angeles.